Materials for High-Temperature Semiconductor Devices

Committee on
Materials for High-Temperature
Semiconductor Devices

National Materials Advisory Board
Commission on Engineering and Technical Systems
National Research Council

NMAB-474
National Academy Press
Washington, D.C. 1995

NOTICE: The project that is the subject of this report was approved by the Governing Board of the National Research Council, whose members are drawn from the councils of the National Academy of Sciences, the National Academy of Engineering, and the Institute of Medicine. The members of the committee responsible for the report were chosen for their special competencies and with regard for appropriate balance.

This report has been reviewed by a group other than the authors according to procedures approved by a Report Review Committee consisting of members of the National Academy of Sciences, the National Academy of Engineering, and the Institute of Medicine.

This study by the National Materials Advisory Board was conducted under ARPA Order No. 8475 issued by DARPA/CMO under Contract No. MDA 972-92-C-0028 with the U.S. Department of Defense and the National Aeronautics and Space Administration.

The views and conclusions contained in this document are those of the authors and should not be interpreted as representing the official policies, either expressed or implied, of the Defense Advanced Research Projects Agency or the U.S. Government.

Library of Congress Catalog Card Number 95-70760
International Standard Book Number 0-309-05335-8

Available in limited supply from:
National Materials Advisory Board
2101 Constitution Avenue, NW
Washington, D.C. 20418
202-334-3505
nmab@nas.edu

Additional copies are available for sale from:
National Academy Press
2101 Constitution Avenue, NW
Box 285
Washington, D.C. 20055
800-624-6242
202-334-3313 (in the Washington Metropolitan Area)

Copyright 1995 by the National Academy of Sciences. All rights reserved.

Printed in the United States of America.

Abstract

Major benefits to system architecture would result if cooling systems for components could be eliminated without compromising performance (e.g., power, efficiency, and speed). The existence of commercially available high-temperature semiconductor devices would be an enabling technology in such areas as sensors and controls for aircraft, high-power switching devices for the electric power industry, and control electronics for the nuclear power industry. This report surveys the state of the art for the three major wide bandgap materials for high-temperature semiconductor devices (i.e., silicon carbide, the nitrides, and diamond); assesses the national and international efforts to develop high-temperature semiconductors; identifies the technical barriers to their development and manufacture; determines the criteria for successfully packaging and integrating new high-temperature semiconductors into existing systems; recommends future research priorities; and suggests additional possible applications and advantages.

The National Academy of Sciences is a private, nonprofit, self-perpetuating society of distinguished scholars engaged in scientific and engineering research, dedicated to the furtherance of science and technology and to their use for the general welfare. Upon the authority of the charter granted to it by the Congress in 1863, the Academy has a mandate that requires it to advise the federal government on scientific and technical matters. Dr. Bruce Alberts is president of the National Academy of Sciences.

The National Academy of Engineering was established in 1964, under the charter of the National Academy of Sciences, as a parallel organization of outstanding engineers. It is autonomous in its administration and in the selection of its members, sharing with the National Academy of Sciences the responsibility for advising the federal government. The National Academy of Engineering also sponsors engineering programs aimed at meeting national needs, encourages education and research, and recognizes the superior achievements of engineers. Dr. Harold Liebowitz is president of the National Academy of Engineering.

The Institute of Medicine was established in 1970 by the National Academy of Sciences to secure the services of eminent members of appropriate professions in the examination of policy matters pertaining to the health of the public. The Institute acts under the responsibility given to the National Academy of Sciences by its congressional charter to be an advisor to the federal government and, upon its own initiative, to identify issues of medical care, research, and education. Dr. Kenneth I. Shine is president of the Institute of Medicine.

The National Research Council was organized by the National Academy of Sciences in 1916 to associate the broad community of science and technology with the Academy's purposes of furthering knowledge and advising the federal government. Functioning in accordance with general policies determined by the Academy, the Council has become the principal operating agency of both the National Academy of Sciences and the National Academy of Engineering in providing services to the government, the public, and the scientific and engineering communities. The Council is administered jointly by both Academies and the Institute of Medicine. Dr. Bruce Alberts and Dr. Harold Liebowitz are chairman and vice chairman, respectively, of the National Research Council.

Committee on Materials for High-Temperature Semiconductor Devices

WOLFGANG J. CHOYKE, *Chair*, Professor, Department of Physics, University of Pittsburgh, Pennsylvania
MICHAEL G. ADLERSTEIN, Consulting Scientist, Raytheon Research Division, Lexington, Massachusetts
JEROME J. CUOMO, Professor. Materials Science and Engineering, North Carolina State University, Raleigh
ARTHUR G. FOYT, Jr., Manager, Electronics Research, United Technologies Research Center, East Hartford, Connecticut
EVELYN L. HU, Chair, Department of Electrical and Computer Engineering, University of California, Santa Barbara
LIONEL C. KIMERLING, Professor, Materials Science and Engineering, Massachusetts Institute of Technology, Cambridge
MARK R. PINTO, Department Head, ULSI, AT&T Bell Laboratories, Murray Hill, New Jersey
MICHAEL A. TAMOR, Staff Scientist, Ford Motor Company, Dearborn, Michigan
IWONA TURLIK, Vice-President, Corporate Manufacturing Research Center, Motorola, Schaumburg, Illinois

LIAISON REPRESENTATIVES

JANE A. ALEXANDER, ARPA/MTO, Arlington, Virginia
T.J. ALLARD, Sandia National Laboratories, Albuquerque, New Mexico
DON KING, Sandia National Laboratories, Albuquerque, New Mexico
WILLIAM C. MITCHEL, U.S. Air Force, Wright Patterson Air Force Base, Ohio
YOON SOO PARK, Office of Naval Research, Arlington, Virginia
J. ANTHONY POWELL, NASA Lewis Center, Cleveland, Ohio
JOHN PRATER, Army Research Office, Research Triangle Park, North Carolina
MAX YODER, Office of Naval Research, Arlington, Virginia

NMAB STAFF

ROBERT M. EHRENREICH, Senior Program Manager
PAT WILLIAMS, Senior Secretary

National Materials Advisory Board

JAMES C. WILLIAMS, *Chair*, General Electric Company, Cincinnati, Ohio
JAN D. ACHENBACH, Northwestern University, Evanston, Illinois
BILL R. APPLETON, Oak Ridge National Laboratory, Oak Ridge, Tennessee
ROBERT R. BEEBE, Tucson, Arizona
I. MELVIN BERNSTEIN, Tufts University, Medford, Massachusetts
J. KEITH BRIMACOMBE, University of British Columbia, Vancouver, Canada
JOHN V. BUSCH, IBIS Associates, Inc., Wellesley, Massachusetts
HARRY E. COOK, University of Illinois, Urbana
ROBERT EAGAN, Sandia National Laboratories, Albuquerque, New Mexico
CAROLYN HANSSON, Queen's University, Kingston, Ontario, Canada
KRISTINA M. JOHNSON, University of Colorado, Boulder
LIONEL C. KIMERLING, Massachusetts Institute of Technology, Cambridge
JAMES E. MCGRATH, Virginia Polytechnic Institute and State University, Blacksburg
RICHARD S. MULLER, University of California, Berkeley
ELSA REICHMANIS, AT&T Bell Laboratories, Murray Hill, New Jersey
EDGAR A. STARKE, University of Virginia, Charlottesville
JOHN STRINGER, Electric Power Research Institute, Palo Alto, California
KATHLEEN C. TAYLOR, General Motors Corporation, Warren, Michigan
JAMES WAGNER, Johns Hopkins University, Baltimore, Maryland
JOSEPH WIRTH, Raychem Corporation, Menlo Park, California

ROBERT E. SCHAFRIK, Director

Preface

Just as human operators must be protected from extreme environments, so must the electronics that operate and control a functional system. When the environment proves too warm, the electronics must be insulated, refrigerated, or simply moved to cooler locations. This last option is sometimes very difficult, or impossible, and the perceived fragility of electronics must then be reconsidered. Vacuum-tube technology provides a historical example of this process. Although vacuum tubes may be considered mechanically fragile, tube-based radio proximity fuses were nevertheless incorporated into artillery shells over 50 years ago! More recently, well-logging electronics derived from available semiconductor technology have been forced to operate for prolonged periods at 300 °C, far exceeding the "standard" limit of 125 °C that appears on uncounted specification documents. When the requirement is unavoidable and the motivation is high (e.g., commercial or military advantage), "accepted" temperature limits need not be accepted.

There is a huge difference between what can be done in principle and what should be undertaken in practice, however. If the question were to be asked "if a family of proven high-temperature electronics functions (for the moment meaning anything higher than 125 °C) were suddenly to become available, would its ultimate economic value justify the cost of its development," the answer is likely to be YES. *This position is further strengthened by the fact that the shared virtues of radiation hardness, power handling, and blue-light emission represent an important leverage for the development of high-temperature semiconductors.* However, if the question were "is there already a market for high-temperature electronics sufficient to justify development of all or part of the family function," the answer may not be so clear. In all the processes of our economy, there are currently few in which insertion of electronics into such environments is absolutely required to achieve acceptable functionality. Recognizing that a human operator can usually be protected and that a central controlling computer is easier still to protect, the determination of whether the benefits of high-temperature electronics will justify the cost requires the examination of how products and processes might be improved, or even enabled, by high-temperature electronics.

The use of distributed control network architectures and embedded processors is rapidly growing. In a crude biological analogy, an animal is more agile, efficient, and durable when its nervous system (sensor signal processing), skeleton (physical structure), and muscular system (actuator operation) are integrated. Electronics are integrated into systems for several reasons: (1) to simplify control paths, thereby simplifying wiring complexity, reducing weight, and improving reliability; (2) to distribute control, allowing robust system reliability and system architecture simplification; (3) to permit operational information to be gathered and processed with greater speed, accuracy, and reliability; and (4) to control actuators. For systems that encounter or generate high temperatures, this integration, or entwining, demands that electronics work at, or near, their functional temperature limit; the more intimate the integration, the greater the environmental stress.

If the economic value of extended-temperature electronics justify its cost, a natural question arises: "since the possibility of high-temperature electronics has been known for decades and the need is so great, why wasn't this done some time ago?" Although there can be no definitive answer to this question, there have been two historical barriers to the development of high-temperature electronics.

First, the functions and performance goals of most familiar complex electronic systems (e.g., telecommunications and computers) are defined and measured in purely electronic terms. Thus, although it can be elaborate and expensive, the need for heat protection is viewed as an unavoidable element of system design, rather than of function.

Second, nonelectronic systems (e.g., turbine engines, nuclear reactors, chemical refineries, and metallurgical mills) are operable without embedded electronic systems. Since the electronic function is not the defining element of these systems and extended-temperature electronics are not available as a robust off-the-shelf technology, many prospective customers will not usually consider such systems. Thus, although cognizant of the architectural advantages of high-temperature electronics, prospective developers have not perceived a general *commercial* market sufficient to justify aggressive development.

Even with these barriers, however, considerable international resources are currently being devoted to developing electronic technologies either tailored for or supportive of high-temperature operation. There is a divergence in the central emphases of these efforts.

- **United States**—Much of the focus is on high-temperature electronics. One manufacturer markets a family of silicon-based integrated circuits suitable for prolonged operation at 250 °C, derived in part from radiation-hardened technologies developed for military applications. Silicon-carbide-based devices are being developed for some control applications and rudimentary diamond-based devices have been demonstrated. Radiation-hardened electronics for reactor control and waste monitoring are avidly sought in both the United States and Europe. The large bandgap and smaller neutron cross sections of the lighter elements in high-temperature semiconductors also translate to radiation damage resistance. There are approximately seven industry, three university, and two national laboratory programs currently active in the high-temperature semiconductor field. The committee was briefed by representatives of most of these programs, which are listed on pages vi and vii. There is also some funding of wide bandgap semiconductors for use in high-power devices (e.g., the Semiconductor Research Corporation program at Purdue University).
- **Europe**—Effort is mainly focused on power electronics. This is synergistic with high temperature because the generation of internal heat is a limiting factor in power devices and is mitigated by larger bandgap and higher thermal conductivity materials. A collaborative organization, HITEN, was formed in 1992 to coordinate European nascent efforts in high-temperature electronics.
  - **Sweden**—Approximately 55 people are engaged in research at Linkoping University and Kista in Stockholm. This is a joint government-ABB industries effort on power electronics, the first goal of which is a 12 kV thyristor.
  - **Germany**—The Deutsche Forschungs Gemeinschaft (DFG) sponsors several universities with Interdisciplinary Research Grants for silicon carbide (SiC). Primary among these are the University of Erlangen-Nürnberg and the Friedrich Schiller University in Jena, which are concentrating on novel growth techniques and electrical and optical measurements. Siemens Research Laboratories in Erlangen, however, are concen-

trating on power devices, as is Daimler-Benz in Frankfurt for electric cars. These laboratories as well as several collaborating universities (i.e., Regensburg, Erlangen-Nürnberg, TH Aachen, Ilmenau, and Fraunhofer Institut fur Angewandte Festkörperphysik in Freiburg) have large BMFT contracts for the development of SiC power-devices.
- **France**—At least 10 university laboratories as well as LETI-Grenoble and Thomson CSF (Paris) have government funding for SiC high-frequency and other devices.

- **Japan**—The committee was unable to discover critical details about the industrial involvement of Japanese companies in SiC development. However, emphasis appears to be on optoelectronics with occasional mention of high-temperature applications for the automotive and aerospace industries. Optical data transmission rates and storage densities are enhanced by the use of shorter wavelength laser light, which is synergistic with high-temperature work because it requires larger bandgap semiconductors. However, the 1994 domestic Japanese SiC conference drew 160 participants, many of whom were interested in power devices. In the nitrides (i.e., gallium nitride, gallium-indium-nitride) light sources, Nichia Chemical is producing a 3 percent efficient blue light-emitting diode. The interest in Japan in large bandgap semiconductors for opto-electronics purposes is highly visible, but an interest for power electronics is growing. Japanese universities that are active in SiC are the University of Kyoto, the Kyoto Institute of Technology, Osaka University, and the Electrotechnical Laboratory in Tsukuba. Nitrides research is also being pursued at Nagoya University.

Against this assessment of the national and international efforts to develop high-temperature semiconductors, the goals of this study are to (1) identify the technical barriers to the development and manufacture of high-temperature semiconductor materials; (2) determine the criteria for successfully packaging and integrating new high-temperature semiconductors into existing systems; (3) recommend future research priorities; and (4) suggest additional possible applications and advantages.

The report is structured as follows. Chapter 1 discusses the need for high-temperature electronics. Chapter 2 reviews the state of the art of wide bandgap materials. The fundamental limit to high-temperature operation is the energy of the semiconducting bandgap of the host material. By this measure, even silicon with its "small" bandgap (1.1 eV) is not widely used near its limit of 300 °C (silicon as a high-temperature material is discussed in Appendix A). Although the technology has not been optimized for high temperature and there are concerns about its chemical stability, gallium arsenide (1.4 eV) does offer the prospect of significantly higher temperature in a mature technology (gallium arsenide is discussed as a high-temperature semiconductor in Appendix B). Alternative materials for yet higher temperatures must be selected with care; larger gap is necessary but not sufficient. Sulfide semiconductors have large bandgaps but decompose at high temperatures. Thus, Chapter 2 reviews the state of the art of materials alternatives for which the prospect of robust high-temperature operation has been confirmed. These include SiC (2.4-3.3 eV depending on polytype), gallium nitride (3.5 eV), aluminum nitride (6.2 eV), boron-nitride (>6.4 eV), and diamond (5.4 eV). Chapters 3-6 discuss generic, technological issues related to the design, fabrication, packaging, and testing of high-temperature circuits and devices (specific case-studies are presented in Appendix C). These chapters contain common elements that must be established for any high-temperature electronics technology to be possible. Chapter 7 presents recommendations as to how to overcome critical hurdles on the path to a family of robust high-temperature electronic devices.

Acknowledgements

The committee expresses its appreciation to the following individuals for their presentations to the committee: Dr. H.M. Hobgood, Westinghouse Science and Technical Center, Pittsburgh; Dr. Calvin Carter, Jr., CREE Research Incorporated, Durham, North Carolina; Professor Peter Barnes, Auburn University; Mr. R.C. Clarke, Westinghouse Science and Technology Center, Pittsburgh; Dr. Joseph S. Shor, Kulite Semiconductor, Leonia, New Jersey; Dr. John Palmour, CREE Research Incorporated, Durham, North Carolina.; Dr. Dale M. Brown, General Electric, Schenectady, New York; Professor Robert J. Trew, Case Western Reserve University, Cleveland; Dr. Terrance Lee Aselage, Sandia National Laboratory, Albuquerque; Dr. Michael W. Geis, Lincoln Laboratory, Massachusetts Institute of Technology; Dr. Jeff Glass, North Carolina State University, Raleigh; Dr. Asif Khan, APA Optics, Inc., Blaine, Minnesota; Dr. Gary McGuire, Center for Microelectronic System Technologies, MCNC, Research Triangle Park, North Carolina; Professor Hadis Morkoc, University of Illinois-Urbana; Dr. Nate Newman, University of California, Berkeley; Dr. Harold West, Honeywell, Incorporated, Plymouth, Minnesota; Dr. Gerald Witt, AFOSR/NE, Bolling Air Force Base, Washington, D.C.; Professor Manijeh Razeghi, Director, Center for Quantum Devices, Northwestern University; Dr. John A. Spitznagel, Westinghouse Science and Technology Center, Pittsburgh; Professor Aris Christou, Chairman, Department of Materials and Nuclear Engineering, University of Maryland, College Park; Dr. Richard Eden, Consultant, Thousand Oaks, California; Professor R. Wayne Johnson, Electrical Engineering Department, Auburn University; and Dr. Philip L. Dreike, Sandia National Laboratory, Albuquerque.

The committee acknowledges with thanks the contributions of Robert M. Ehrenreich, Senior Program Manager; Jack Hughes, Research Associate; and Pat Williams, Senior Secretary, to the project.

Contents

EXECUTIVE SUMMARY		1
1	BACKGROUND	7
	Survey I: Applications of High-Temperature Electronics by Industry, 7	
	Survey II: Applications by Thermal Environment, 12	
	Survey III: High-Temperature Electronics Applications by Complexity, 13	
	Summary, 14	
2	STATE OF THE ART OF WIDE BANDGAP MATERIALS	15
	Silicon Carbide, 15	
	Nitride Materials, 24	
	Diamond, 28	
3	DEVICE PHYSICS: BEHAVIOR AT ELEVATED TEMPERATURES	31
	High-Temperature Effects: Fundamental, Materials-Related Properties, 31	
	Predicting High-Temperature-Device Performance: Materials-Related Figures of Merit, 33	
4	GENERIC TECHNICAL ISSUES ASSOCIATED WITH MATERIALS FOR HIGH-TEMPERATURE SEMICONDUCTORS	39
	Electrical Contacts, 39	
	Doping and Implantation, 40	
	Gate Oxides and Insulators, 43	
	Etching, 45	
	Defect Engineering and Control, 46	
	Yield, 47	
	Device reliability, 48	
5	HIGH-TEMPERATURE ELECTRONIC PACKAGING	51
	Chip Packaging, 51	
	Substrates, 53	
	Thick-Film and Thin-Film Metallization, 53	
	Component Attachment, 55	
	Interconnection, 56	
	Second-Level Packaging, 57	
	Summary, 58	

6   DEVICE TESTING FOR HIGH-TEMPERATURE ELECTRONIC MATERIALS   61
    Short-Term Constant-Temperature Tests, 61
    Constant-Temperature Life Tests, 62
    Thermal-Cycling Tests, 62
    Future Requirements for High-Temperature Testing, 63

7   CONCLUSIONS AND RECOMMENDATIONS   65
    General Conclusions and Recommendations, 65
    Materials-Specific Conclusions and Recommendations, 67

References   71

Appendix A: Silicon as a High-Temperature Material   81

Appendix B: Gallium Arsenide as a High-Temperature Material   87

Appendix C: High-Temperature Microwave Devices   93

Appendix D: Biographical Sketches of Committee Members   119

Figures

1-1	Schematic of a hypothetical drive-by-wire system for an automobile with computerized traction control, steering, and suspension	8
1-2	Log-log plot of the complexity of some example applications as a function of temperature	13
2-1	Average values of the optical constants of SiC from the vacuum ultraviolet to the middle infrared	17
2-2	Calculated band structure of 3C-SiC	18
2-3	Calculated band structure of 2H-SiC	18
2-4	Summary of the experimentally observed exciton bandgaps and their temperature variation for the different SiC polytypes	19
2-5	Thermal conductivity of two single crystals of SiC	20
2-6	Schematic showing the basic elements of the modified sublimation process	22
2-7	Schematic of a typical SiC CVD growth chamber	23
2-8	Band structure of hexagonal and cubic modifications of AlN	26
2-9	Band structure of hexagonal and cubic modifications of GaN	27
2-10	Band-structure calculation of diamond	29
2-11	Thermal conductivity of two Type IIa diamonds	29
3-1	Calculated electron mobility as a function of temperature for undoped 6H-SiC and 3C-SiC	32
3-2	Calculated electron mobility as a function of temperature for GaN doped n-type, 10^{17} cm^{-3}	32
3-3	Intrinsic carrier density for silicon, GaAs, and SiC	32
3-4	Decrease in silicon bandgap with increasing temperature	33
3-5	Calculated reverse leakage current densities in p-n junctions of various materials	35
3-6	Variation in threshold voltage versus temperature for n- and p-channel MOSFET devices	36
3-7	Operating temperatures for different devices per material	37
4-1	Schematic of the device structure for a AlN/Al$_x$Ga$_{1-x}$N SISFET	44
4-2	Increase in resistivity of unintentionally doped Al$_x$Ga$_{1-x}$N with increasing aluminum mole fraction	45
5-1	Decrease in insulation resistance as a function of temperature	52
5-2	Reduction from nine to three electrical path segments between two integrated circuits with multichip module technology	59
6-1	Variations in threshold voltage for p- and n-type silicon MOSFETs with temperature	61
6-2	Drain characteristics of a SiC inversion-mode MOSFET at 650 °C	61
A-1	Reduction in large junction isolation areas by the use of trenches and SOI	83
A-2	Leakage currents as function of temperature for three types of n-MOS transistors with gate lengths of 2 microns	83

A-3	Schematics of the dielectric isolation material process flow and the bonded wafer material process flow	84
A-4	Open-loop gain as a function of temperature	85
B-1	GaAs MESFET and silicon MOSFET drain leakage currents	88
B-2	MESFET transconductance, g_m, after three-hour anneals at various temperatures	88
B-3	Diffusion barrier constructed of nine alternating layers of electron-beam evaporated tungsten and silicon	89
B-4	Comparison of conventional MESFET with MESFET using temperature-hard ohmic contacts, buried p-type channel implants, and gate sidewall spacers	89
B-5	MESFET showing on/off current ratio decreasing from 106:1 at room temperature to near 20:1 at 400°C	90
B-6	High-temperature MESFET incorporating modifications to standard process	90
B-7	Operating characteristics of MESFET structure shown in Figure B-6	91
C-1	Contours of normalized power dissipation on the gain-efficiency plane	94
C-2	Enhancement- and depletion-mode MOSFETs	96
C-3	Structure of a bipolar junction transistor	97
C-4	Simulated microwave performance of SiC BJTs	98
C-5	Comparison of SIT with MESFET: (a) potential gate barriers established, (b) resulting current-voltage curves for SIT; (c) generic MESFET I-V curves	98
C-6	Structure of the Junction Field Effect Transistor (JFET)	100
C-7	Typical current-voltage curves for a JFET at various temperatures	101
C-8	Structure of an inverted JFET in SiC	101
C-9	Measured small signal current and unilateral gain for SiC MESFETs	103
C-10	IMPATT diode performance compared with projections for wide bandgap semiconductors	104
C-11	Material structures and electric field profiles possible for IMPATT diodes	104
C-12	A simplified equivalent circuit for an IMPATT diode embedded in a microwave circuit	105
C-13	'a' contours for MESFETs of silicon, GaAs, silicon carbide, and gallium nitride	106
C-14	Calculated locus of drain-current saturation for (a) silicon carbide, (b) silicon, and (c) GaAs (with and without parasitic series resistance)	107
C-15	A simple model for ohmic contact and channel resistance contributions to MESFET source resistance	108
C-16	Contact resistance calculated as a function of contact length for three materials	109
C-17	Contours of constant Z plotted on r_c-R_{sq} plane	110
C-18	Representation of current-voltage curves for a MESFET and typical loadlines for Class A operation	111
C-19	Small signal equivalent circuit for a MESFET	112
C-20	Contours of constant temperature rise in the GaAs MESFET channel	113
C-21	Contours of constant temperature rise in the SiC MESFET channel	114
C-22	A MODFET transistor with a two-dimensional electron gas at the interface between GaN and AlGaN	115

Tables

2-1	Comparison of Semiconductor Properties	16
2-2	Notations for Selected SiC Polytypes	17
2-3	Exciton Binding, Nitrogen Ionization, and Valley-Orbit Splitting Energies and Effective Mass for SiC Polytypes	21
3-1	Comparison of Normalized Figures of Merit of Various Semiconductors for High-Power and High-Frequency Unipolar Devices	34
4-1	Selected Ohmic Contacts to n-Type 6H-SiC and Measured Contact Resistivities at Room Temperature	40
4-2	Selected Ohmic Contacts to p-Type 6H-SiC and Measured Contact Resistivities at Room Temperature	41
4-3	Additional Ohmic Contact for SiC	42
4-4	Ohmic Contacts for GaN	43
5-1	Properties of Ceramic AlN, Ceramic SiC, Glass $\pm$ Ceramics as Compared with 90 percent Alumina	54
5-2	Metallizations for AlN Substrates	55
5-3	Dielectrics for AlN Substrates	56
5-4	Summary of Properties of Metallizations for AlN	57
5-5	Typical Cofired Metals	58
6-1	Short-Term Constant-Temperature Tests	62
6-2	Constant-Temperature Life Tests	63
C-1	Summary of Room-Temperature DC Gain for Various Field Effect Transistors of SiC	102
C-2	Assumed and Calculated MESFET Current-Voltage Model Parameters	108
C-3	Listing of Several Refractory Metallizations on SiC and their Contact Resistivities	111
C-4	Assumed and Calculated MESFET Power Model Parameters	112

Executive Summary

Electronics that operate and control functional systems must currently be protected from extreme environments. Major benefits to system architecture would result if cooling systems for electronic components could be eliminated without compromising system performance (e.g., power, efficiency, speed). The existence of *commercially available* high-temperature semiconductor devices would provide significant benefits in such areas as:

- sensors and controls for automobiles and aircraft;
- high-power switching devices for the electric power industry, electric vehicles, etc.; and
- control electronics for the nuclear power industry.

With the possible exception of light-emitting diodes (LEDs), however, present commercial demand for wide bandgap semiconductor materials is limited. While there are few pressing applications that cannot be achieved without wide bandgap materials, the vast array of applications, and hence, the value, will only be realized once these materials have evolved to such an extent that off-the-shelf devices are available.

At the request of the U.S. Department of Defense and the National Aeronautics and Space Administration, the National Materials Advisory Board of the National Research Council convened the Committee on Materials for High-Temperature Semiconductor Devices to assess the national and international efforts to develop high-temperature semiconductors; to identify the technical barriers to their development and manufacture; to determine the criteria for successfully packaging and integrating new high-temperature semiconductors into existing systems; to recommend future research priorities; and to suggest additional, possible applications and advantages.

This Executive Summary is divided into two sections. The first section presents general conclusions and recommendations about future research priorities to accelerate the acceptance of high-temperature semiconductor materials. This section discusses the temperature ranges for the different materials to be used, the competitiveness of U.S. research versus foreign competition, the systems in which high-temperature electronic materials should initially be introduced, and the government/industry/university collaborations required to forward the development of high-temperature semiconductor materials. The second section discusses the barriers to the successful development, manufacture, packaging, and integration of wide bandgap materials into existing systems and presents the key research and development priorities to overcome these barriers.

GENERAL CONCLUSIONS AND RECOMMENDATIONS

Temperature Ranges

Silicon and silicon-on-insulator electronics may be sufficient for some applications for temperatures up to 300 °C. Such applications include digital logic, some memory technologies, and some derated analog and power applications. Silicon-based technology will not be sufficient for many applications operating in the 200-300 °C range, however, such as power-conditioning devices in higher-temperature control systems. These devices will have to be produced from another material system. *Based on the evidence presented in this report, silicon-carbide-based devices are currently in the best position to meet this need, particularly n-channel enhancement-mode metal-oxide semiconductor field effect* transistors (MOSFETs). *However, significant technological barriers, such as micropipes, oxide quality, contacts, metallization, packaging, and reliability evaluation still need to be further addressed.*

As a result of fundamental limitations, silicon-based technologies will not be useful at temperatures above 300 °C. Other materials must be used for these temperature ranges, but the choices are somewhat less clear. Technology based on gallium arsenide (GaAs) might be used for systems operating up to 400 °C. Just working at elevated temperatures is not the only concern, however. It is also essential that the devices reliably function over a wide range from very cold (i.e., -20 °C) to very hot (i.e., 400 °C). *Based on the evidence presented in this report, devices based on n-type silicon carbide (SiC) are the only type that currently appear to meet the temperature-range and reliability requirements, but additional development is needed.* Eventually, high-temperature electronic technology could be developed for reliable operation even for temperatures above 600 °C.

U.S. Competitiveness

As described in the Preface, considerable international resources are currently being devoted to developing electronic technologies either tailored for or supportive of high-temperature operation. The United States is focusing most of its efforts on high-temperature applications and currently has a slight lead in SiC research.

Europe appears to be increasing its effort in wide bandgap materials, especially for power electronics. This research area is synergistic with high-temperature applications because the generation of internal heat is a limiting factor in power devices and can be mitigated by larger bandgap and higher thermal conductivity materials. The dedication of European resources to this area is seen in the founding of the collaborative organization HITEN, which was established in 1992 to coordinate nascent European efforts in high-temperature electronics.

Japan is emphasizing the use of wide bandgap materials for opto-electronics and leads in the use of nitrides for light sources. Japan is also becoming interested in power and high-temperature applications. Unfortunately, the closed nature of Japanese industry made it difficult for the committee to determine the true level of interest in wide bandgap materials research. The increased interest in high-power, high-temperature applications is evident in Japan's annual domestic SiC conference, however. The Third Domestic (Japan) SiC Conference convened in Osaka on October 27-28, 1994, with approximately 160 experts in attendance. Contrary to Japan's previous two conferences, there was a greater emphasis at the Osaka conference on high-power, high-temperature applications than on LEDs.

The Commonwealth of Independent States had a number of major programs in SiC development, but the current financial difficulties of most of the Commonwealth's institutions are preventing many laboratories from continuing their research. There is a wealth of expertise and information available for leveraging by other countries, however. For instance, the European Community is planning on supporting a SiC growth effort in St. Petersburg (Y.M. Tairov and V.E. Chelnekov, personal communication, 1994).

The committee believes that the U.S. wide bandgap materials research community is currently very competitive in the international research community. *To remain competitive in the international research community, the committee recommends that demonstration technologies be pursued to motivate further research and increase interest in high-temperature semiconductor applications.*

Demonstration Technologies

To increase interest and motivate further research in wide bandgap materials, a realistic, inspiring application focus must be found that can make system designers aware of the benefits of high-temperature electronics. A wide bandgap transistor that operates at 150 °C will not drive the technology because it will be in direct competition with the more economically efficient silicon technologies. The demonstration technologies must be *system* circuits (i.e., not an *individual* device) that can be inserted into essentially nonelectronic systems (e.g., turbine engine, nuclear reactor, chemical refinery, or metallurgical mill) with the goal of measurably increasing system performance.

As discussed in Chapter 1, the committee believes that there eventually will be a niche market for semiconductors with temperature capabilities higher than that of silicon, and that this market will be sufficiently large to justify the cost of development. However, this belief is tempered by the recognition that because such electronics will be used in new ways there is little immediate demand. The market will grow only in synergy with the availability of components. This suggests that development of high-temperature electronics not be undertaken in isolation. Instead, such development can and should be leveraged from development of other technologies with more immediate applications, thus reducing the costs and risks of both. Three suitable application areas are high-power electronics, nuclear reactor electronics, and opto-electronics.

Power switching devices, for example, would be a good demonstration technology for high-temperature semiconductor materials. High-voltage, high-power electronics, while not necessarily used as high-temperature devices, nevertheless need wide bandgap semiconductors because of their superior breakdown voltages and high thermal conductivities. There is already considerable research being pursued in this area because (1) improved high-power switching devices could save an estimated $6 billion in the cost of construction of additional transmission lines; and (2) the smoother, more efficient use of the transmission system would reduce the need for new generating capacity, which the Electric Power Research Institute estimates would be a savings of $50 billion in North America alone over the next 25 years (Spitznagel, 1994).

The pursuit of demonstration technologies would not only increase interest in wide bandgap materials, it would also provide significant test beds for the application of the technology and enhance our understanding of the *generic* technologies required to further high-temperature-device operation (e.g., materials etching and implantation; degradation modes of metallic gates, contacts, and interconnects at high temperatures; packaging behavior at high temperatures; and accelerated-testing and reliability-testing methodologies to ensure proper functioning). *The ability to grow a reasonably defect-free material is not the only requirement for the realization of a successful technology. The development of demonstration technologies would also help identify other factors that must be resolved for high-temperature electronics to be incorporated into existing systems.*

Funding Strategy

The need for new development funds for demonstration technologies and future wide bandgap materials is not necessary in the committee's opinion. Government funding currently exists for long-range research in wide bandgap materials, although additional funding would certainly allow more options to be evaluated within a shorter period of time. Industry has also demonstrated a willingness to commercialize new developments if the projected payback to their investments can occur within the short term (NRC, 1993). *The committee believes that the high-temperature research community should leverage the research funding for wide bandgap materials that is currently being provided by the high-power and optics markets, where no viable alternatives to wide bandgap materials currently exist.* Building on the funding for other areas dependant on wide bandgap materials reduces the need for potential users of high-temperature devices to fund the required materials development exclusively and, thus, may render it cost effective.

The committee recommends the following strategy for the development of wide bandgap materials:

- develop precompetitive alliances and integrated programs (national laboratories, universities, and industries) for coordinating research, technical skills, and capabilities to expedite research in the most efficient manner;
- direct research at a technology demonstrator that has definite applications (i.e., is a product) and addresses the usually neglected areas of packaging, assembly, testing, and reliability (e.g., high-power switches; integrated motor control; power phase shifter);
- concurrently develop materials, design, testing, and packaging; and
- build and test the demonstration component on a cost-share basis that encourages teaming, ensures adequate funds, and requires periodic deliveries.

The committee believes that the founding of a newsletter that provides a summary of published worldwide developments in high-temperature semiconductor research would assist the establishment, development, and maintenance of (1) a fundamental long-term materials effort, (2) an infrastructure within the industry, (3) a group to monitor international development, and (4) a U.S. information group for highlighting advances.

MATERIALS-SPECIFIC CONCLUSIONS AND RECOMMENDATIONS

The first three parts of this section concentrate on the major wide bandgap materials discussed in this report: SiC, nitrides, and diamond. The final part of this section concerns the generic problems in packaging that will affect the production of all high-temperature electronic devices.

Silicon Carbide

SiC is an indirect bandgap semiconductor and has enjoyed the longest history and greatest development with regard to both materials growth and device realization. As such, SiC is currently the most advanced of the wide bandgap semiconductor materials and in the best position for near-term commercial application. Its main application will be in high-power, high-temperature, high-frequency, and high-radiation environments. It will not be suitable for blue lasers or ultraviolet light emitters, however, except as a potential substrate material. The three key research efforts for the development of commercially viable SiC devices are:

- *Wafer production:* The 1- and 2- inch SiC wafers now in production are rapidly approaching *device quality* where they might be used for commercial production of devices and circuits with acceptable yield. It could be argued that such small wafers are entirely sufficient for what will be a relatively small market (compared with silicon) with a very high-price premium, and therefore an early investment in larger wafers is not justified. However, the entire commercial infrastructure for electronics manufacture is based on a wafer size of at least 3 inches, and preferably 4 inches, as a minimum. Reconstructing a small-wafer infrastructure that became obsolete over 30 years ago will be both an expense and an obstacle to the introduction of commercial SiC electronics. The committee believes that the development of larger SiC wafers is viewed as the more cost-effective approach to commercial development.
- *Film growth:* Chemical vapor deposition, molecular-beam epitaxy, and other film-growth technologies and chemistries require refinement to produce epitaxial films with n- and p-type doping ranges from 10^{13} to 10^{20} cm^{-3} for nitrogen, aluminum, boron, gallium, transition metals, and rare earth elements.
- *Manufacturing processes:* Lower-cost device-production methods are required to make the manufacture of SiC devices more competitive with the silicon technologies.

Nitrides

Interest in the direct bandgap nitride materials (i.e., gallium nitride, aluminum nitride, aluminum gallium nitride, and indium gallium nitride) has dramatically increased recently because of their optical properties. The materials show great promise and are likely to dominate the visible and ultraviolet opto-electronics market. Nichia's recent bright blue LEDs have already stimulated increased industrial effort (e.g., Hewlett Packard, Spectra Diode Laboratories, Xerox PARC) in materials growth, contact metallurgy and reliability, and device reliability and testing, although the materials have defect densities of greater than 10^{10}/cm^2 and the mechanism of photo emission is currently unknown. Heterojunctions in the nitrides also hold promise for higher-speed devices compared with SiC. Their applicability for power development and high-frequency devices is unproven at this time, and the technologies for wafer production, doping, and etching are currently less developed than SiC and require more longer-term research before they will be competitive with other electronic materials. However, as development of photonic applications for wide bandgap materials progresses, the opto-electronic market may provide an effective way to leverage the development of these materials for high-temperature-device applications. The committee identified the following three research efforts as being key to the development of nitride devices:

- *Compatible substrates:* Better-matched substrates are required for nitride wafer production to be commercially tenable.
- *Wafer production:* Growth of quasi-crystalline films of gallium nitride, aluminum gallium nitride, and aluminum nitride should be pursued on substrates such as SiC to gain thermal advantages.

- *Doping:* Methods for both n- and p-type doping of Group III nitrides are required.

Diamond

Diamond is a well-understood material, but its use for active electronic device applications is not feasible at this time because of the difficulties associated with its economical growth and doping. While diamond transistors have been designed, fabricated, and tested, their performance is also orders of magnitude less than that which is expected from the electrical properties intrinsic to diamond. The poor performance is thought to result from excessive nitrogen impurities and from as yet not fully explained surface-depletion effects. The current prognosis for diamond is primarily as a protective coating, a thermal management film, and a material for electron-emitting cathodes.

Packaging

Much more research is required in the area of high-temperature packaging. For high-temperature electronics to be commercially viable and provide true performance advantages, interconnection and packaging technologies are required that can reliably operate at temperatures up to 600 °C for 10^4 hours. To attain these goals, innovative packaging techniques will be required. The three key research efforts for the development of high-temperature packages are:

- *Metallization:* Contacts are required in the 10^{-6} to 10^{-7} Ω/cm^2 range that have long-term durability at temperatures up to 600 °C. Greater understanding is needed of the long-term effects of high temperatures on contact and interconnect metallurgy, degradation and failure modes, reliability, and interfaces.
- *Device reliability and aging testing:* Existing methods of accelerated, environmental-life testing of packages must be adapted for high-temperature applications to ensure the accurate assessment of device reliability and aging.
- *Computer-aided design tools:* Computer-aided design tools are required that incorporate electrical and mechanical simulation of high-temperature electronic systems.

1

Background

Trying to enumerate systematically all the possible applications for new high-temperature electronics would be a futile endeavor. Rarely are all of the conceivable uses for any new technology obvious. The committee was able to identify only a few eager potential users that currently have active programs that require higher-temperature electronics. Several more applications are under consideration but are not in active development. Any group of technologists could generate a much larger list of plausible applications. However, while these applications might seem reasonable to enthusiasts of high-temperature electronics, they may not be realistic options to the prospective customers for this new technology. Furthermore, just as for the microprocessor (or any number of other new technologies, such as the laser), a much larger array of "enabled" applications is likely to evolve if, and when, a proven off-the-shelf technology becomes a viable option in engineering new products and systems. Rather than generating one more speculative list, a few of the better-defined applications are described in this chapter, supplemented with more generic descriptions of environments and applications for high-temperature electronics.

The largest possible range of applications can be anticipated by means of three surveys. The first survey is a traditional list describing applications ranging from programs in progress, through speculative system designs, to what amounts to a few responses to the question: What might be done differently if *cost-effective* high-temperature electronics were available? A systematic estimate of the economic value of high-temperature electronics was not attempted by this committee, but expert estimates that are available are included in this survey. The second survey classifies the types of environment that might be encountered by electronics and then associates some of the previously identified applications with each environmental type. In principle, a general classification of all possible operating environments would automatically describe the environments associated with all possible applications, including those not yet conceived. The third survey again uses the list of identified applications to give a sense of the capabilities that might be needed as a function of temperature. Although the first survey is, by definition, incomplete and the second and third are hardly more than intellectual exercises, together they give a strong sense of the potential industrial and economic importance of high-temperature electronics.

SURVEY I: APPLICATIONS OF HIGH-TEMPERATURE ELECTRONICS BY INDUSTRY

Automotive

The automotive industry is often cited as the primary near-term market for high-temperature electronics. While the automotive environment is stressful to electronic systems, the stress is rarely in the form of simple heat. Conventional vehicle architectures with an open-bottomed front engine compartment, generous underhood and underbody airflows, a metal heat-dissipating body and frame structure, and access to a water-cooling circuit leave very few locations within a vehicle that regularly achieve temperatures significantly above 100 °C. These locations are mainly near the exhaust system or brakes and can usually be avoided. Occasional problems with reliability due to high temperatures (as high as 150 °C) have been addressed by combinations of heat shielding, redirected airflow, blowers, or simple component relocation. Except for rare cases of architectural errors, the major challenge to reliability of automotive electronics is the combination of rapidly changing environmental stresses (temperature and humidity cycling), exposure to corrosives and solvents, and an economic mandate for low-cost packaging. With careful attention to device and circuit layout, wire-bond and lead-frame integrity, choice and use of polymer packaging materials, and strict process control, automotive electronics actually meet or exceed military specifications at a small fraction of the cost and in huge volumes (Motz and Vincent, 1984; Dell'Acqua and Marelli, 1990; Frank and Valentine, 1990).

Despite the illusion of a comfortable status quo, four trends are forcing major changes in the approach to automotive electronic component and system design. First, even with current vehicle architectures, customer expectations of reliability continue to rise. Flawless performance for 10 years or 150,000 miles will soon be standard. Second, the electronics content of modern automobiles is rising rapidly, both in convenience features (e.g., heads-up display and navigation systems) and in the management of powertrain and suspension systems. Figure 1-1 is a diagram of a hypothetical drive-by-wire system with computerized traction control, steering, and suspension. The amount of sensing, signal processing, data transfer, system control, and power actuation is very large. A few elements of this system (e.g., semi-active suspension, antilock brakes, and traction control) are currently in the marketplace. Multiplex wiring will soon be standard in motor vehicles. While easing the transfer of information and reducing wiring weight and complexity, multiplex wiring dictates the location of quite complex nodes in many hostile locations. Third, the physical architecture of the vehicle itself is changing. Improved aerodynamics dictates more compact flowing shapes with less internal airflow, which forces denser packaging of the powertrain and exhaust systems. Serious consideration is being given to sealing the engine compartment and moving the radiator to the rear of the vehicle. Fourth, replacement of metal body and frame components with composites of much lower thermal conductivity will eliminate many safe havens for electronics. This is not so much a trend to higher temperature as a trend toward more uniform temperature; locations near 100 °C may disappear while those between 150 °C and 200 °C will remain plentiful. Nevertheless, solutions that evolved for the current, more open, steel-based architecture may not serve in the hotter environments of future vehicles.

Power electronics are also rapidly proliferating in automobiles (Thornton, 1992; Bose, 1993). Figure 1-1 indicates several systems that include high-power actuators. Full, active suspension requires several tens of kilowatts. Electric and hybrid-electric vehicles are totally dependent on power electronics for efficient operation of motor and braking systems. There are two types of high-temperature issues for power electronics.

First, in conventional combustion-powered vehicles, the electronics must be placed somewhere that is preferably near or within the device they control. Safe, cool locations have become scarce, however. For example, the drivers for electric, active front-suspension components share the underhood environment, while the flywheel-mounted motor and alternator for torque leveling are cooled only by the engine oil and may reach 300 °C! For electric and hybrid vehicles, the wiring weight, resistive losses, and radio frequency emissions are minimized by placing the power electronics within the motor housing. To minimize weight, these motors are sized such that they may produce several times their continuous-service power for periods of several seconds. This translates to a rapid temperature rise that is currently limited to 180 °C only by the magnetic properties of the permanent magnet rotor. Integral power electronics must survive repeated and rapid excursions to at least this temperature.

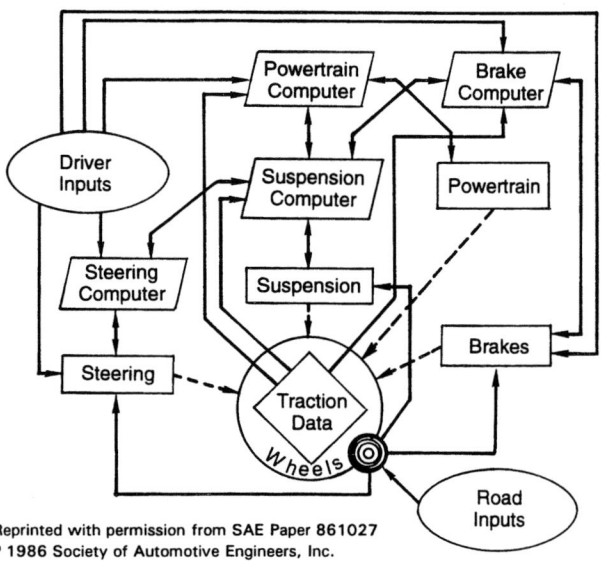

Reprinted with permission from SAE Paper 861027
© 1986 Society of Automotive Engineers, Inc.

FIGURE 1-1 Schematic of a hypothetical drive-by-wire system for an automobile with computerized traction control, steering, and suspension. SOURCE: Rivard (1986).

Second, power devices generate considerable internal heat. This heat must be dissipated to prevent thermal runaway (i.e., when heat generated increases with temperature) and destructive failure. In many locations, a cool-sink for large amounts of heat is unavailable. Although active cooling is always an option, a power-device technology immune to thermal runaway is highly desirable. The smaller package size afforded by a higher-temperature technology is of considerable value on its own in terms of thermal management.

In summary, two clear needs can be identified for automotive electronics. First is the need for a low-cost, highly reliable technology for operation at an intermediate temperature (perhaps 200 °C). This need might be served by modification of current silicon-based technology. Second is the need for power electronics able to function in elevated ambient temperatures with restricted heat-sinking. As current silicon-based power technology is largely limited by internal heat generation, a switch to a wide bandgap semiconductor is dictated.

Aerospace

Gas Turbine Engines

High-temperature electronics are essential to the development of multiplexed systems for gas turbine engine control (Nieberding and Powell, 1982). In present control systems, all electronics are centralized in a protected area that is cooled with ambient air or fuel. This architecture has proven satisfactory for some time, but, as the requirements for engine control become increasingly complex, the wire harness and connectors associated with point-to-point architecture have become major weight and reliability issues. Some wire harnesses weigh over 150 pounds, and every connector is a point of system vulnerability. A solution to this problem is to introduce a multiplexed architecture in which wire harnesses are replaced with common busses, a change that demands high-temperature electronics.

The Air Force Integrated High-Performance Turbine Engine Technology Program is a multiphase project aimed at achieving increased thrust, 50 percent weight reduction, fault-tolerant control, and system integration of military aircraft engines. A key element in this program is the development of higher-temperature electronics. The environments for electronics in an aircraft engine cover a wide range for some potential sensors: 175-800 °C. Early phases of the project call for electronics and optics for operation at 175 °C, while an intermediate phase calls for 250 °C. The final phase of the program anticipates *heat-sink* temperatures as high as 350 °C. Specifications for commercial engines are not yet available but are likely to be similar (Skira and Agnello, 1992; Tillman and Ikeler, 1992). Some of these temperatures are suitable for devices based on silicon technology, while others lie beyond that currently anticipated for any electronics technology. While it is neither desirable nor cost effective (and maybe impossible) to construct the whole system to survive the highest temperatures, any increase in operating temperature offers a corresponding increase in design flexibility.

Other Aerospace Applications

Engines demand the highest temperature requirements for current aircraft, but temperature requirements will rise in many other critical areas as vehicle speed increases. A recent example is the control of the engine inlet guide vane for the high-speed civil transport, which requires that the moderately complex electronics driving the guide vane actuators operate for prolonged periods at 200 °C.

In high-performance or heavily electronics-laden aircraft (today almost the same thing), a generic problem appears: as speed—and therefore heat generation—and altitude increase, the ability to dissipate waste heat into the atmosphere decreases (Christenson, 1991). Locations in the aircraft that remain below 125 °C or that can be conveniently reached by the cooling system cannot be found. Many electronics systems, including avionics, radars, and communications equipment, must be derated in performance to maintain even the minimal acceptable reliability at the margins of their operating ranges. Fuel is often used as the medium for heat transfer within the aircraft, but some fuel must then be kept in reserve as essentially dead weight and, when cooling to outside air is insufficient, the fuel tanks become a limited heat-sink. The cooling techniques currently in use force tradeoffs between speed, altitude, and systems shutdown.

In one example of a supersonic fighter plane, 90 percent of the cooling capacity of the environmental control system (ECS) is devoted to electronics and only 10 percent to the pilot! The single ECS unit weighs roughly 2,000 lbs and consumes 50 kW of power. Its excessive weight precludes the addition of a backup, so its failure aborts the mission. A smaller ECS, possibly with a backup, would reduce weight and power while increasing overall system reliability. *Higher-temperature electronics will enhance reliability and enable major changes in the electronics architecture of aircraft.*

Space Vehicles and Exploration

Problems directly related to high temperature are rare once in space; space is cold and intense sunlight may be reflected with high efficiency. There are several situations in which high temperature may be an issue, however. First, sensing and control of rocket boosters and thrusters may require proximity to the hot plumbing associated with combustion. Such problems and issues are very similar to those for aircraft jet engines, with the notable exception that maintainability and long-term reliability are less important. Second, some space exploration vehicles must enter hot environments. A proposed balloon-borne probe of Venus' atmosphere must operate at 325 °C, while a Venus lander must endure 460 °C. Closer approaches to Mercury or the sun would also require higher-temperature electronics. Third, material and design factors that support high-temperature electronics operation would also enhance radiation hardness and increase resistance to upsets and damage from the unavoidable flux of cosmic radiation (Jurgens, 1982).

Nuclear Power

There are two types of nuclear power applications for wide bandgap semiconductors: those associated with reactor operation and those associated with handling, processing, and storing of radioactive waste. It has been reported that material and devices in high-temperature operations tend to be resistant to radiation damage (Knoll, 1989). The highest temperature reached in a properly operating pressurized water reactor (PWR) is nearly 300 °C. Although this temperature is actually somewhat lower than that used in combustion power plants, accessibility is much more limited and difficulty of repair or replacement demands much higher reliability.

High-temperature, radiation-hard electronics can improve PWR operation by improving reactor control and reducing expensive and occasionally hazardous repairs. The most important areas relate to monitoring and control over the distribution of power generation in the core of the reactor. At present, a three-dimensional map of the core is developed from an array of thermocouples and neutron-flux detectors distributed through the reactor core. These require numerous penetrations (roughly 60) of the reactor vessel and must be replaced every three years. With integrated-drive electronics and multiplexing, a different detector type would last at least twice as long and require only four penetrations. *By the year 2010, this alteration would amount to a savings of nearly half a billion dollars in materials and over $100 million in avoided costs of radiation exposure for the 100 operating PWRs in the United States* (Spitznagel, 1994).

With the limit on penetrations relieved, more detectors might be used to provide a more detailed "map" of the core. In-core measurement of the water level also enhances this mapping. A more accurate map of the core allows for safer operation, more efficient consumption of the fuel, and extension of the period between shutdowns for refueling. Downtime, whether deliberate or forced, has been conservatively estimated to cost roughly $500,000 per day (NRC, 1993). At an estimated $50,000 per man-rem of radiation exposure, repairs and maintenance are very expensive (Spitznagel, 1994). *If radiation-hard, high-temperature-electronics control and monitoring devices could improve the current "nuclear generating capacity factor" from 65 percent to a reasonable target of 85 percent, then yearly savings per plant would be $36.5 million per year. With at least 100 plants in the United States, this constitutes a yearly savings of $3.6 billion per year.* This is an impressive savings and should be a great encouragement to the development of high-temperature semiconductors.

Other PWR applications include monitoring of boron and nitrogen 16 in the water. The thousands of valves and pipes in a reactor must be monitored for proper valve positioning, corrosion, and fatigue. Following an accident, the environment of the reactor containment building can be hot (420 °C), wet, and radioactive. Actuators and sensors must survive under these conditions.

High-temperature electronics may also play a role in radioactive waste storage and handling. The condition of stored nuclear waste must be monitored. Buildup of explosive gasses must be prevented, as must leaks of toxic or radioactive material. This requires sensors in both the tanks and the surrounding environment. Monitors include neutron and gamma radiation monitors, temperature sensors, chemical sensors (gas and liquid), leak sensors, and cameras. Currently, conventional television cameras survive approximately only 30 minutes in a nuclear storage container (Spitznagel, 1994). A radiation-hard television camera would be a great asset in reactor monitoring, repair, and waste handling. Under an accident condition described above wherein the containment building may become hot, wet, and radioactive, remote visual inspection of a damaged reactor is extremely difficult with current technology. Robust monitoring equipment will also limit the need for opening the containers for maintenance.

The case of the orbiting power reactor (e.g., the Russian-designed Topaz) combines all the difficulties of nuclear power with space electronics. Although space is indeed cold, heat is only lost by radiation. The size of the radiator for the cold end of a heat engine (i.e., the reactor) increases very rapidly as the cold-end temperature is reduced. Raising this temperature allows a much more compact design but exposes the control electronics to higher temperatures. At the same time, radiation shielding for sensitive electronics is wasted payload weight. High-temperature, radiation-hard electronics would allow a smaller, lighter, and simpler design for a space-borne reactor.

Petroleum Exploration

Well-logging is a strong driver for high-temperature electronics. Modern petroleum exploration involves elaborate probing of wells during drilling. For this reason, oil exploration companies have been some of the earliest customers for high-temperature electronics. Earlier efforts have resulted in fairly complex circuits built of discrete devices that are able to operate for periods of several hundred hours at temperatures up to 300 °C. Since it is very expensive to withdraw and replace probes during drilling, reliability is of extreme importance. An off-the-shelf family of more sophisticated components would enable far more reliable and effective logging tools.

Industrial Process Control

Industrial process control is rarely mentioned in the context of high-temperature electronics, but may prove to be one of the most important high-temperature electronics applications. Most monitoring of high-temperature industrial processes (e.g., refining, annealing, baking, and curing) involves monitoring the process flow from a fixed location. While this monitoring may expose some sensors to temperature extremes and other hazards, the associated electronics are easily protected and cooled. Some processes are best observed from inside, however. For example, careful control of the time-temperature profile during epoxy curing is a key element to yield and reliability in the electronics industry. Appropriately insulated recorders and transmitters are currently sent through baking and curing ovens, but these devices are expensive. In this example, cure temperatures do not exceed 200 °C, and many others do not exceed 300 °C. Thus, these temperatures are within reach of many high-temperature technologies, which would offer the possibility of widely available, inexpensive, compact sensors, memories, and transponders that ride through the high-temperature process beside, or even buried within, the product. It may even be possible to report the temperature and stresses on an integrated circuit while the package is being formed or to attach coded identifiers to components that record and report on their history throughout the manufacturing process. Such "smart tags" would be useful for process and quality control (Arbab et al., 1993).

Power Electronics

The importance of power electronics in vehicles was discussed earlier. Many of the issues concerning internally generated heat and in-motor integration also apply to many other applications. In vehicles, there are three general areas of application. These include high-torque induction-motor controllers, high-efficiency voltage converters and switches, and variable high-voltage ultra-capacitors (Miller, 1987). The integration of control and power electronics—so-called "smart power"—is certainly an architectural advantage. There are many additional applications for small, high-torque electric motors besides motor vehicles. Such small motors will replace hydraulics in many applications once the reliability issues are settled, offering considerable design and control advantages and eliminating the weight of hydraulic fluid and the complexity of associated plumbing.

A good example of extensive integration of power electronics is the Air Force's More-Electric Airplane. The anticipated advantages of all-electric actuation are considerable: a 20-30 percent reduction in system weight and cost; fivefold increase in system efficiency; fivefold reduction in heat generation; faster system response; and improved maintainability, reliability, and survivability. The More-Electric Airplane incorporates a starter/generator as an integral part of each engine. The generator will provide all auxiliary (nonthrust) power for aircraft operation. In current aircraft, a smaller generator is connected by a gear shaft. Temperatures in the new location already exceed 125 °C and may exceed 200 °C in future engines. Furthermore, the very high powers involved (hundreds of kW) dictate that the power-conditioning electronics be located close to the generator and the engine. Thus, both cooling and "remoting" are not attractive options and high-temperature electronics are highly desirable. There are other military "More-Electric" programs that relate to armored vehicles, ships, submarines, and even the individual combat soldier. Similar civilian "electric-hydraulic" applications include lighter and more agile industrial robots and more precise and efficient excavation and earth-moving equipment.

Electric power was once measured simply by its cost and quantity. Recently, the quality of electricity has become a serious issue. Disturbances to line voltage and noise on power lines is disruptive to such sensitive systems as computers. Utility power conditioning has been identified as a key area for application of power electronics on a large scale (Hingorani and Stahlkopf, 1993). On average, roughly a third of the rated capacity of the power transmission grid is unused in the United States. This margin is held to absorb very large inductance transients from disturbances (e.g., generator failure, overload cutouts, and broken cables) without damaging switching and generating equipment. Large-scale power electronics would allow real-time phase-shifting of utility power and provide this protection while allowing nearly 100 percent use of the national power grid. *The Electric Power Research Institute (EPRI) estimates an available savings of $6 billion compared to the cost of additional transmission lines of the same capacity. Smoother and more efficient use of the transmission system also reduces the need for spare generating capacity. EPRI estimates that this efficient use would create a savings of $50 billion in North America alone over the next 25 years.* With higher-quality power available directly from the utility grid, the need for uninterruptable power supplies will be greatly reduced. As these are at best only 80 percent efficient, their elimination would effectively increase power-generating capacity at essentially no cost.

SURVEY II: APPLICATIONS BY THERMAL ENVIRONMENT

Three factors define the thermal environment for electronics: (1) ambient temperature, to which a quiescent device will inevitably rise in the absence of any circulating coolant; (2) external temperature gradients around the device or module, which are defined by the details of the nature of the application; and (3) internal temperature gradients, which are generated by *active* devices. When these factors appear singly, high-temperature applications can be classified as immersion (i.e., no temperature gradients and therefore no cold-sink to cool the devices), proximity (i.e., the application brings the electronics close to a hot region but does not dictate immersion; at least a limited cold-sink is available), and internal (i.e., where internally generated heat must be removed to a cold-sink). Temperatures for these applications are discussed in the next section.

Examples of purely immersion applications include reactor monitoring, well-logging, ride-through process monitoring, some nodes in aircraft or motor vehicle multiplex systems, and the Venus lander. In such applications, every component of the system must perform satisfactorily at the nominal operating temperature. An example is combustion-flame sensing for jet engine control. The sensor itself must survive a very hot location with line-of-sight to the combustion chamber while the associated interface signal circuit is placed as close as possible. Obviously, there are design and cost tradeoffs in how much of the system needs to be exposed to the nominal high-temperature environment. Support electronics may be removed to cooler locations at the expense of cabling and reduced signal.

Proximity applications typically appear where some high-temperature component or process must be monitored or where system architecture motivates incorporation of control electronics near a very hot component. An example is the engine-mounted control computer for automobiles. While exposing the computer to the increased temperatures associated with the large *gradients* of the exhaust system, moving the computer from the vehicle to the engine realizes two advantages. It allows calibration of the computer to the specific engine on which it is mounted (rather than a single-model engine) for improved performance and reduced emissions. It also minimizes the number of wires connecting the engine to the vehicle, simplifying assembly and improving reliability.

Generally, internal heating is a major issue only for power electronics. Power electronics must be incorporated wherever electrical actuation is required. To a first approximation, heat generated by power devices simply superimposes an internally generated gradient on the externally defined thermal environment and raises the nominal-device operating temperature accordingly. Power devices appear in both immersion and proximity applications. Examples of "immersed" power electronics are the torque-leveling motor and integrated traction motor described in the previous section. A case of power devices in proximity to a hot region would appear in any case where the actuating motor grows extremely hot or the objects to be actuated are hot themselves. Such situations will appear in many aircraft and vehicle control applications (e.g., the inlet guide vane mentioned earlier). Just as for nonpower proximity, device temperature may be reduced at the expense of system integration.

One important consideration is that the temperature rise in a power device can be very large. For silicon-based power devices, junction temperatures in excess of 200 °C are apt to result in catastrophic failure. Current systems are engineered to sustain the rated power output with heat-sinking into a 100 °C ambient, which is adequate to keep the devices below their failure temperature; in effect, they are designed to operate at the edge of disaster. The same rate of heat generation and heat-sinking capacity into a 200 °C ambient (or cold-sink temperature) would imply a junction temperature of 300 °C, beyond the abilities of silicon. This simplistic linear analysis suggests that even a small increase in ambient temperature for silicon-based power electronics will require a combination of improvements in heat extraction and derating of the devices themselves. Both of these changes increase the size and cost of the systems. The heat extraction problem is further compounded by the fact that thermal conductivity of most materials decreases with increasing temperature. Power electronics based on wider bandgap semiconductors would address this issue.

SURVEY III: HIGH-TEMPERATURE ELECTRONICS APPLICATIONS BY COMPLEXITY

The ability to satisfy the need for electronics for a given temperature is predominantly a function of what is required for the application. Complexity, as crudely measured by the number of active devices in the module or system node, varies by nearly seven orders of magnitude. Figure 1-2 is a log-log plot of the complexity of some of the applications identified in Survey I as a function of their temperature. Because the scales are logarithmic, large errors in either parameter cannot eliminate the obvious trend. This figure suggests three general categories of high-temperature electronics.

The first category includes all the complexity and functionality now available in conventional silicon technology that is functional to roughly 200 °C (e.g., memories, microprocessors, analog circuits). These applications might be served by modifications of current junction isolation, integrated circuit technology with new metallization and packaging, or if necessary, by silicon-on-insulator technology for operation up to 300 °C. Decreases in device speed and noise margins must be accepted but might be mitigated by changes in device geometry and layout rules. In all the categories discussed,

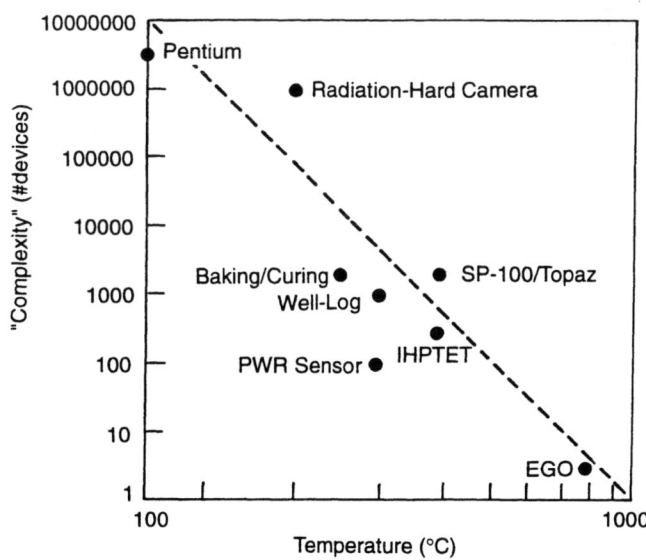

EGO: emission gas-oxygen sensor
IHPTET: integrated high-performance turbine engine technology

FIGURE 1-2 Log-log plot of the complexity of some example applications as a function of temperature.

it must also be remembered that even seemingly minor modifications in technology must be carried out with adequate circuit yield, which is critical at these levels of integration.

The second group includes applications of intermediate complexity, perhaps several dozen to several thousand devices, requiring operation at temperatures of up to roughly 450 °C. This level of complexity is sufficient to support the local functionality of sensing and measurement along with the signal conditioning, basic signal processing and control, limited memory, and interface (via wire or radio) to higher-level systems in cooler environments. Although this definition remains vague, it does appear that no reasonable application calls for duplication of all silicon capabilities in a 450 °C technology. It does appear that a more limited family of devices, integrated circuits, and circuit-board technology will be necessary for these applications.

The third group of applications generally involve sensing of one or more parameters of a very hot environment. Examples include automobile exhaust-gas analysis and jet engine flame detection, which are considered proximity applications; plausible immersion applications above 500 °C have not been identified. In these two cases, the sensor design is driven by its function and the required environment. For the automotive exhaust gas-oxygen sensor, this is 700-900 °C in a strongly oxidizing or reducing atmosphere. EGO sensors use TiO_2 or ZnO_2 as wide bandgap semiconductors in which oxygen ions behave as holes. Thus, these high-temperature applications involve only one, or at most a few, active devices: the sensor itself and the minimal biasing and correction circuitry. The very large temperature gradients (several hundred degrees centigrade in a few centimeters) in most proximity applications could be made to appear inside the module. This allows use of intermediate-range electronics in support of the high-temperature sensing component.

One element omitted from this temperature-complexity scatter plot is internally generated heat from power devices. They can be treated as individual "hot" devices in need of lower-temperature support electronics, analogous to the high-temperature sensing applications. The critical difference is that the temperatures are much lower. While silicon power devices may run into difficulties in ambient temperatures much above 100 °C, the low-power support electronics could easily be made to function at much higher temperatures. This strongly suggests a mixed technology consisting of silicon-based control electronics from the first category in support of power devices in a wide bandgap semiconductor technology.

SUMMARY

Although it is impossible to anticipate all possible applications for high-temperature electronics, it is possible to categorize them. A real need exists for advanced microprocessors functional to 200 °C, but system complexity appears to decrease rapidly with required operating temperatures. Thus, some natural groupings appear that suggest directions for technological development. The low-temperature, high-complexity applications require a silicon-based technology modified for reliable operation up to at least 200 °C, with a reduced family of functions at 300 °C. Intermediate-complexity, intermediate-temperature applications require rudimentary integrated circuit technology (i.e., logic functions, small memories, and analog signal devices) and discrete circuit technology for circuits containing several dozen to several thousand devices operational to 450 °C. Low-complexity, high-temperature applications are driven by sensing. A family of devices for such high temperatures is probably not necessary. The sensors themselves are per force designed for such environments, and a slightly more benign environment suitable for intermediate temperature devices is usually quite near. The need for reduced package size and weight and higher operating temperature defines a pressing need for wide bandgap power devices.

2

State of the Art of Wide Bandgap Materials

This chapter surveys the state of the art for the three major wide bandgap materials for high-temperature semiconductor devices: silicon carbide (SiC), the nitrides, and diamond. This chapter is not a comprehensive examination of all the properties of the different materials, but does examine closely those properties related to high-temperature operation. The intrinsic properties of the wide bandgap materials versus those of the more common silicon and gallium arsenide (GaAs) materials are compared in Table 2-1.[1] Although silicon and GaAs are not considered in this chapter because of their expected limited high-temperature applicability, devices and interconnects of these materials are discussed in Appendices A and B, respectively.

SILICON CARBIDE

Materials Description and Properties

Of the wide bandgap materials, SiC is by far the most developed. The earliest reported recognition of the silicon-carbon (Si-C) bond is by Berzelius in 1824. SiC has been produced in the United States since 1891 when Eugene G. Acheson (1893) of Monongahela City, Pennsylvania, melted a mass of carbon and aluminum silicate by passing a current through a carbon rod immersed in the mixture. A variety of vapor-transport furnaces have been used in this century to grow boules of single-crystal SiC. In addition, high-purity homo-epitaxial single-crystal films of SiC have been grown in both horizontal and vertical chemical vapor deposition (CVD) reactors.

Moisson reported in 1904 and 1905 that hexagonal crystals of SiC were present in meteoritic specimens from Canyon Diablo, Arizona. Naturally occurring SiC was viewed as exclusively of extraterrestrial origin until 1957. However, SiC has recently been discovered in alluvial sands and in Kimberlite breccia in a number of locations on the earth.

SiC forms in a variety of crystal structures, termed polytypes, of which over 175 have been described in the literature (Verma and Krishna, 1966; Pandy and Krishna, 1983). Only simple polytypes are of interest for SiC devices. Their basic crystallographic stacking sequences and most common notations are illustrated in Table 2-2 (Verma and Krishna, 1966). The optical properties of SiC do not differ very much from polytype to polytype (Figure 2-1).

To better understand SiC, a brief discussion of electronic band structure is warranted. Band-structure calculations for SiC have been made for the past 30 years, but theorists have concentrated on the zincblende 3C-SiC polytype and the wurtzite 2H-SiC structure since the other polytypes are much more complicated due to their much larger unit cells. The accuracy of such calculations has recently been considerably improved and currently there is a sizable effort to work on the band structures of 4H-, 6H-, and 15R-SiC. Early band-structure calculations of 3C and 2H are shown in Figures 2-2 and 2-3 to provide a qualitative "feel of the neighborhood" where the maxima in the valence band and the minima in the conduction band are likely to be located. Since both 3C-SiC and 2H-SiC are *indirect-gap semiconductors*, it is reasonable to assume that all polytypes are indirect-gap semiconductors. Indeed, experiment has verified that in addition to 3C- and 2H-SiC, 4H-, 6H-, 8H-, 15R-, 21R-, 27R-, and 33R-SiC are also indirect-gap semiconductors. Figure 2-4 summarizes the experimentally observed exciton bandgaps

---

[1] Table 2-1 was developed for comparative purposes using the data that was available during the course of this study. This table should not be considered a definitive tabulation of the properties of these materials, since new, more accurate data are constantly being accumulated for most of these materials.

TABLE 2.1 Comparison of Semiconductor Properties

Properties	Silicon	GaAs	3C-SiC	4H-SiC	6H-SiC	GaN	Diamond	AlN
Lattice constant (Å) (RT)	5.430	5.65	4.3596	3.073 10.053	3.0806 15.1173	4.51	3.567	3.11 a_0 4.98 c_0
Thermal expansion (x 10^{-6}) °C	2.6	5.9	4.7	—	4.2 a_0 4.7 c_0	5.6	0.08	4.5
Density (g/cm^3)	2.328	5.32	3.210	—	3.211	—	3.515	3.255
Melting point (°C)	1420	1238	2830^a	2830^a	2830^a	—	4000	3000
Bandgap (eV)	1.1	1.43	2.39	3.26	3.02	3.45	5.45	6.2
Saturated electron velocity (x 10^7 cm/s)	1.0	1.0	2.2	2.0	2.0	2.2	2.7	—
Carrier mobility (cm^2/V·s) Electron Hole	1,500 600	8,500 400	1,000 50	1,000 50	370 90	1,250 250	2,200 1,600	14
Breakdown (x 10^5 V/cm)	6	6	20	30	32	>50	100	>50
Dielectric constant	11.8	12.5	9.7	—	9.6-10	11	5.5	10
Resistivity (Ω·cm)	10^3	10^8	—	—	—	10^{10}	10^{13}	10^{13}
Thermal conductivity (W/cm·K)	1.5	0.46	5	4.9	4.9	1.3	20	3.0
Absorption edge (μm)	1.4	0.85	0.50	0.37	0.40	0.36	0.22	0.12
Refractive index	3.5	3.4	2.7	2.712	2.7	—	2.42	3.32
Hardness (kg/mm^2)	1,000	600	3,980	2,130 a_0	—	—	10,000	1,200

^{a} Pressure = 35 bar; decomposes.

NOTE: Dashes indicate information not available.

TABLE 2-2 Notations for Selected SiC Polytypes

Ramsdell Notation	Stacking Sequence	Zhdanov Notation
3C	...ABC...	—
2H	...AB...	11
4H	...ABAC...	22
6H	...ABCBAC...	33
15R	...ABCBABCACBCABAC...	$(32)_3$

and their temperature variation. Experiment has also given an estimate of the binding energy (27 meV) of the exciton in 3C-SiC. Assuming that this value will not be very different in the other polytypes, the actual bandgap, E_G, can be estimated by adding 27 meV to the known value of the exciton bandgap, E_{GX}. Estimates of room-temperature values of both E_G and E_{GX} are given in Figure 2-4. The thermal conductivity is shown in Figure 2-5.

The electrical properties in the various polytypes can be very different because the actual conduction-band minima in the various polytypes will not be in exactly the same positions in the Brillouin Zone. In addition, there is the extra complication of having a different number of nonequivalent sites in different polytypes as a consequence of different size unit cells. This is illustrated for the donor nitrogen in Table 2-3. SiC may be doped n-type with nitrogen up to at least 10^{19} cm^{-3}. The acceptors aluminum and boron can be used to dope SiC p-type to at least 5×10^{18} cm^{-3}. Nitrogen is difficult to keep out of the growth process, and at present unintentional concentrations of nitrogen in the range of 10^{14} cm^{-3} are found in the best epitaxial films. This is sufficiently low not to interfere with current device fabrication.

Deep electronic states due to scandium (Tairov et al., 1974), titanium (Patrick and Choyke, 1974), and vanadium (Maier et al., 1992) have been studied in some detail in various polytypes of SiC. Other deep states, termed D_I and D_{II}, due to implantation or radiation damage have also been widely studied. Many other impurity defect complexes have been observed during annealing of irradiated samples from 0-2000 °C.

Methods of Fabrication

Bulk Growth

The commercial potential of SiC semiconductor technology has been enhanced by recent significant progress in the growth of large single-crystal SiC boules.

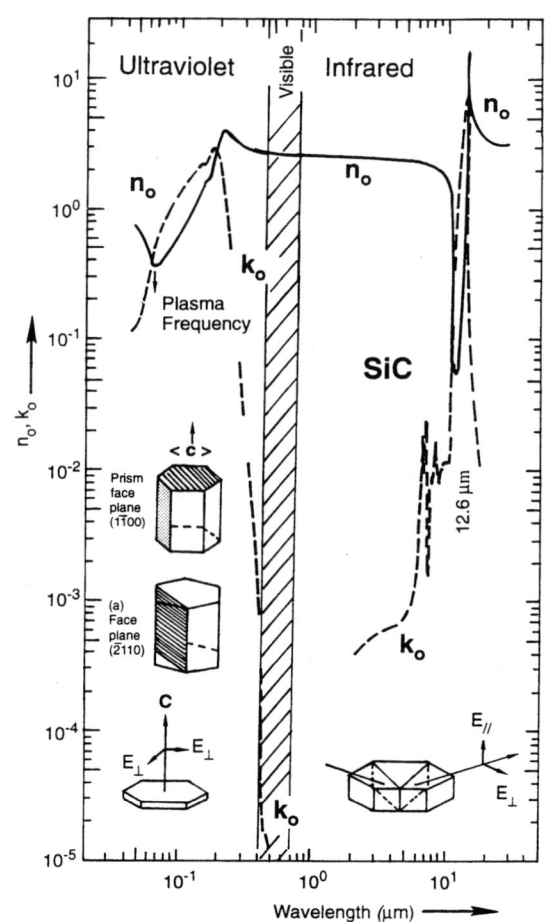

FIGURE 2-1 Average values of the optical constants of SiC from the vacuum ultraviolet to the middle infrared. NOTE: n_o = index of refraction; k_o = extinction coefficient.

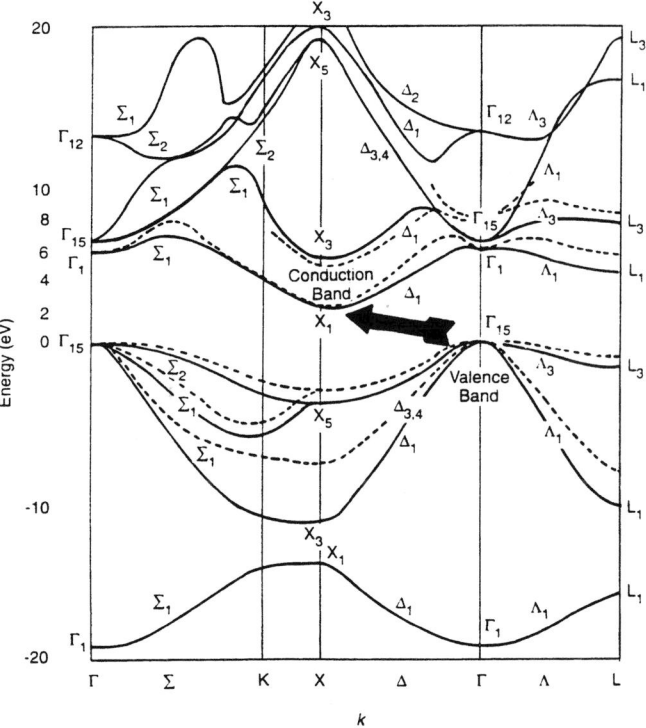

FIGURE 2-2 Calculated band structure of 3C-SiC. SOURCE: Based on Hemstreet and Fong (1974).

For many years, the lack of suitable SiC crystal-growth processes inhibited the commercialization of this promising semiconductor material. There are two properties of SiC that make the growth of bulk single crystals more difficult than that of silicon. First, it does not melt under any reasonably attainable pressure; rather it sublimes at temperatures above 1800 °C. Thus, conventional growth-from-melt techniques (e.g., as for silicon or GaAs) cannot be used for SiC crystal growth. Second, different polytypes with different electronic characteristics can grow under apparently identical conditions (Knippenberg, 1963). A completely satisfactory model for the formation of the various polytypes does not exist. Despite these difficulties, major progress has recently been made in SiC boule growth. The diameter of commercially grown, single-crystal boules is typically 30 mm, and prototype boule diameters have exceeded 50 mm.

Currently, there is interest in at least five of the SiC polytypes: 3C-SiC, 2H-SiC, 4H-SiC, 6H-SiC, and 15R-SiC. Boules of 4H, 6H, and 15R have been grown, and wafers from 4H and 6H boules are commercially available. No significant-sized boules of 3C have been reported. To date, 2H has only been grown in the form of small, millimeter-sized needles.

There are several key review papers that discuss the growth of bulk SiC single crystals (Knippenberg, 1963; Tairov and Tsvetkov, 1983; Powell and Matus, 1989). This section summarizes some of the early work and describes recent developments for which information is publicly available. Much of the current technology is considered to be proprietary and has not been published. Although growth-from-solution techniques have been tried, the most successful growth techniques are based on the sublimation of SiC.

Background. Prior to the mid-1950s, small hexagonally shaped SiC platelet crystals were available through the industrial Acheson process for making abrasive material (Knippenberg, 1963). In 1955, Lely developed a laboratory sublimation process for growing crystals that were much purer (Lely, 1955). In the Lely process, a hollow cavity was formed inside a charge of polycrystalline SiC. The charge was heated to about 2500 °C in a graphite tube furnace at which point the SiC sublimed and condensed on slightly cooler parts of the cavity. Growth took place on a thin, porous graphite cylinder that formed the wall of the cavity. Nucleation was uncontrolled and the resulting crystals were randomly sized, hexagonally shaped α-SiC platelets. These platelets often exhibited a layered structure of various α polytypes. The predominant polytype (generally more than 75

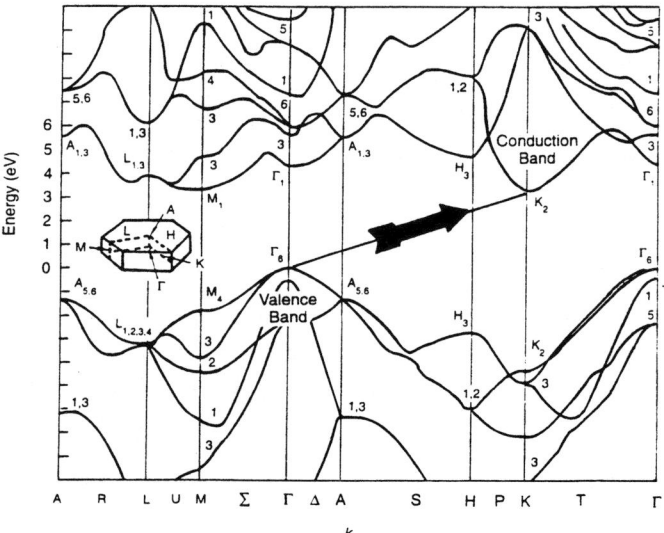

FIGURE 2-3 Calculated band structure of 2H-SiC. SOURCE: Based on Hemstreet and Fong (1974).

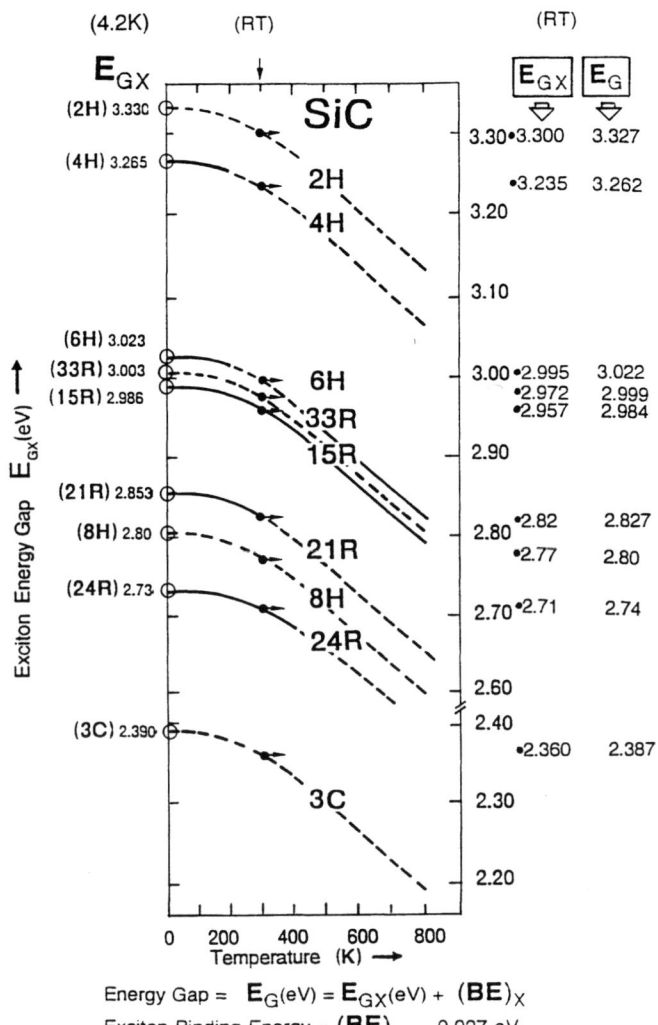

FIGURE 2-4 Summary of the experimentally observed exciton bandgaps and their temperature variation for the different SiC polytypes.

percent) was 6H, followed by 4H and 15R. Although much was learned about SiC from investigations of these crystals over the next 30 years, the process was not suitable for commercial development of SiC.

In the 1970s, Tairov and Tsvetkov (1978, 1981) developed a modification of the Lely process (now commonly called the modified sublimation process, or the modified Lely process) in which growth occurred on a seed crystal. Although some research groups have been somewhat slow in adopting this process, it is now being developed in many labs in Russia, Germany, Japan, and the United States.

The basic elements of the modified sublimation process are shown in Figure 2-6, which is a schematic diagram of the configuration used by Westinghouse. Nucleation takes place on a SiC seed crystal located at one end of a cylindrical cavity. A temperature gradient is established within the cavity such that the polycrystalline SiC is at approximately 2400 °C and the seed crystal is at approximately 2200 °C. At these temperatures and at reduced pressures (argon at 200 Pa), SiC sublimes from the source SiC and condenses on the seed crystal. Growth rates of a few millimeters per hour can be achieved.

Current Status. Cree Research Incorporated of Durham, North Carolina, is the only commercial source in the world of SiC wafers produced from boules. Cree is currently selling 30-mm-diameter wafers of both 4H- and 6H-SiC. Other companies and institutions, known to be producing SiC boules for internal consumption, include Westinghouse, ATM, Siemens, Sanyo, Nippon Steel, Kyoto University, and Kyoto Institute of Technology. Both Cree and Westinghouse have demonstrated boules (and wafers) of up to approximately 50 mm in diameter.

Despite the fact that SiC is extremely hard (between sapphire and diamond in hardness), techniques for cutting and polishing wafers are currently in use. However, the capability is far short of that for silicon. As a result, the polished surface of commercial SiC wafers contains many scratches and defects. Some defects introduced into the wafer by cutting and polishing can be removed by suitable pregrowth (i.e., prior to epitaxy) etching processes (Powell et al., 1991).

Currently, SiC boules (and the commercially available wafers) do contain defects and impurities. One of the most significant defects is a distribution of tubular voids, called micropipes, in the order of a micrometer in diameter (Koga et al., 1992). The micropipes are oriented with respect to their long axis and are approximately parallel to the crystal c-axis; density is typically several hundred per square centimeter. In addition, wafers contain line defects (dislocations) intersecting the surface with a density of 10^4 to 10^5 cm^{-2}. The most common background impurities are nitrogen, aluminum, boron, and metals that can act as deep-level traps.

It has been shown that the micropipes can cause premature reverse breakdown in p-n junctions (Neudeck and Powell, 1994). Evidence shows that microplasmas form in the micropipe at reverse voltages of several hundred volts. The current micropipe density limits the area of high-voltage devices to about 3 mm^2; hence, this defect must be significantly reduced before high-power devices are practical. Several theories have been proposed

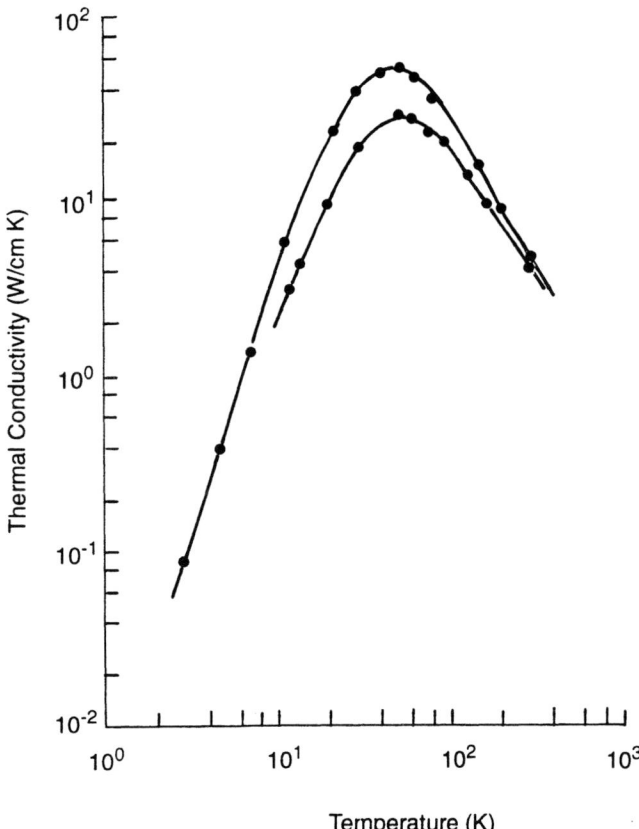

FIGURE 2-5 Thermal conductivity of two single crystals of SiC.
SOURCE: Adapted from Slack (1964).

to explain the formation of micropipes. One theory is based on the presence of contaminant particulates during nucleation and boule growth (Yang, 1993). Another theory is based on the presence of super-screw dislocations (Wang et al., 1993). In this latter theory, hollow cores would form to relieve stress caused by screw dislocations.

Progress is being made in reducing the density of micropipes. In a recent paper, growth of boules in the (1010) direction significantly reduced the formation of micropipes (Takahashi et al., 1994). However, the dislocation density is very high in these crystals. Researchers at Cree have reported (J.W. Palmour, personal communication, 1994) that the density of micropipes has been reduced significantly in the last year. It should be noted that the research team directed by Professor Yu Vodakov at the Ioffe Institute in St. Petersburg, Russia, have produced small single polytype SiC boules (1.5 cm diameter and 7 mm thick) that are claimed to have no micropipes (Y. Vodakov, personal communication, 1994).

Another impediment to wide use of SiC technology is the cost of wafers. At present, there is only one commercial supplier of wafers in the world. The current price per 30-mm-diameter wafer is more than $1,000. This high price can be expected to drop considerably during 1995 as other sources enter the market. The primary reason for this price being lower than GaAs is that both silicon and carbon are 100 times cheaper than gallium.

Epitaxial Growth

Semiconductor-quality α-SiC epitaxial films can now be grown routinely on α-SiC wafers by CVD. In addition, in situ CVD doping processes can produce both n-type and p-type epitaxial films with net carrier concentrations from the 10^{14} cm^{-3} range to greater than 10^{19} cm^{-3}. This technology, which has largely been developed in the last few years, has allowed the development of SiC devices with record-setting performance.

Background. The growth of epitaxial SiC films has many similarities with the growth of epitaxial silicon; however, it has only been recently that the *differences* in growth processes have been appreciated. While conventional semiconductors are grown at approximately two-thirds of their melting temperatures, these temperatures are not practical with wide bandgap materials. For this reason, the substrate temperature cannot be used to assure that all components of the activation energy have been exceeded. In addition, only one crystal structure can be produced in silicon, whereas many crystal structures are possible in SiC. Thus, the polytypic structure of the film must be controlled during formation. The factors that control SiC structure are the crystal orientation and perfection of the substrate. The presence of defects and contamination can also significantly affect the resulting structure. In this report, the term "homo-epitaxy" is used for growth in which the film and substrate are the same polytype, and "hetero-epitaxy" is used when the SiC polytype of the film is different from the substrate. With respect to doping, the incorporation of dopants is dependent on the ratio of the silicon and carbon sources during the growth process and also on the crystal orientation.

TABLE 2-3 Exciton Binding, Nitrogen Ionization, and Valley-Orbit Splitting Energies and Effective Mass for SiC Polytypes

SiC	Exciton Binding to 4D E_{BX} (meV) (PL)	Nitrogen Ionization Energy E_D (meV)				Valley-Orbit Splitting E_{vo} (meV)		Effective Mass $m\perp, m\parallel$ (Cyclotron resonance)
		(Haynes)	(IR)	(2EL)	(Hall)	(IR)	(ERS)	
3C	10	57	—	53.6	20-47	—	8.37	0.247, 0.677
4H	7	47	52.1	—	45	7.6	—	0.42,
	20	96	91.8	—	100	—	—	0.29
6H	16	81	81.0	—	85	12.6	13.0	
	31	136	137.6	—	125		60.3	0.42,
	32	140	142.4	—			62.3	2.0
15R	7	47	49.3	—				
	9	54	59.6	—	53			
	—	—	—	—				
	19	91	—	—	99			
	20	96	—	—				

Techniques used to produce epitaxial SiC films include CVD (Davis et al., 1991), the "sublimation sandwich" process, and liquid-phase epitaxy (Ivanov and Chelnokov, 1992). Homo-epitaxial growth of α-SiC on α-SiC substrates has been achieved by all three techniques. The lack of 3C-SiC substrates has led to a variety of hetero-epitaxial processes to produce 3C-SiC epitaxial films. The 3C-SiC polytype has been grown on silicon, TiC, and α-SiC substrates. These processes are examined in the following sections.

CVD of α-SiC Epitaxial Films. For both α- and 3C-SiC, the CVD process is the current method of choice because, of the three techniques, it yields better films at the lowest temperature. It is also adaptable to commercial production.

A typical SiC CVD growth chamber, shown in Figure 2-7, is similar to chambers used for silicon (Powell et al., 1987). The quartz chamber is water-cooled because growth temperatures are generally higher than those used for silicon epitaxy. The substrates are heated by an inductively heated SiC-coated graphite susceptor. Hydrogen is used as a carrier for various process gases. Prior to growth, the substrates are frequently subjected to an etch with hydrochloric acid (HCl) to reduce defects and contamination. Silane (SiH_4) and propane (C_3H_8) can be used as sources of silicon and carbon during growth. Important system parameters for growth include the growth temperature, flow rates of the various gases, and the silicon/carbon ratio in the gas. Important substrate parameters include the orientation and polarity of the SiC substrate. Typical growth rates are in the 1- to 5- μm/h range. In situ doping is achieved by adding nitrogen or phosphorous for n-type and aluminum (trimethylaluminum, TMA) or boron (diborane) for p-type material. Particular growth and doping processes are discussed.

SiC Epitaxy in the C-axis Direction. An important discovery in SiC epitaxy was that the crystalline orientation of the growth surface is an important growth parameter. In the past, much of the growth was carried out on the "as-grown" (0001) surface (the basal plane) of Lely crystals; that is, growth was in the c-axis direction. The (0001) SiC crystals with polished surfaces have vicinal (0001) orientations, that is, the growth surface may be tilted slightly "off-axis" with respect to the (0001) crystallographic plane. The size of this tilt angle can have

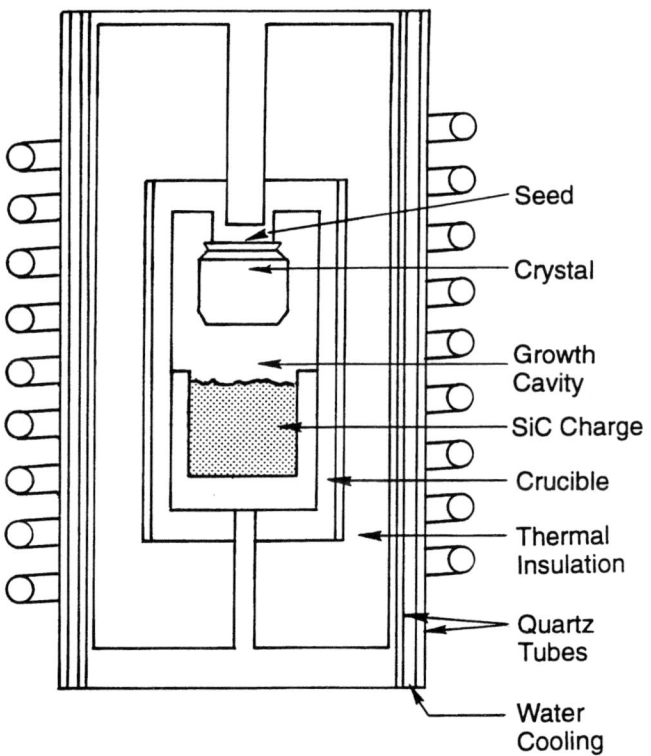

FIGURE 2-6 Schematic showing the basic elements of the modified sublimation process. SOURCE: Hobgood (1993). Courtesy of Westinghouse, Inc.

a dramatic effect on the structure of an epitaxial film. In subsequent discussions in this report, SiC substrates having tilt angles of about 3° are referred to as being "off-axis" and substrates with tilt angles of less than 0.5° are "on-axis." The polarity (i.e., silicon face or carbon face) of the substrate is also an important parameter.

In sublimation sandwich growth, it was found that homo-epitaxy of the various polytypes was enhanced if the growth surface of the substrate was polished off-axis by a few degrees from the (0001) basal plane (Tairov and Tsvetkov, 1983). The research team of Matsunami at Kyoto University discovered that the CVD growth temperature required for producing good-quality 6H-SiC epilayers on 6H-SiC substrates could be reduced from about 1750 °C to about 1450 °C if the growth surface was off-axis by a few degrees from the (0001) plane (Matsunami et al., 1989). They called this growth "step-controlled" epitaxy because growth occurs at steps on the off-axis surface. The stepped surface automatically provides the stacking sequence of the substrate polytype. Hence, homo-epitaxy takes place. 3C-SiC was found to grow at small tilt angles (e.g., less than 1.5°) or at low temperatures because deposited atoms cannot migrate to the steps on large terraces. Also, mobility of deposited atoms is reduced at these lower temperatures and deposited atoms do not reach the steps.

Work at the NASA Lewis Center demonstrated that homo-epitaxy of 6H-SiC on vicinal (0001) 6H-SiC can be achieved at 1450 °C with tilt angles as low as 0.1° (Powell et al., 1991). As a consequence of this result, it was proposed that the cause of the 3C-SiC nucleation was due to defects and contamination on the growth surface. By a suitable pregrowth etching process, the defects and contamination were reduced or eliminated. In effect, there is a competition between defects and surface steps. At sufficiently large tilt angles (high step density), homo-epitaxy will occur even in the presence of defects. At low tilt angles (low step density), any defects that are present will dominate and act as nucleation sites for 3C-SiC. Thus, growth must occur at atomic steps if homo-epitaxy of 6H-SiC is to be achieved. In addition, suitable pregrowth etches can be effective in reducing or eliminating defects caused by cutting and polishing the SiC substrate.

Homo-epitaxial SiC films on vicinal (0001) SiC substrates have been obtained with the 4H-, 6H-, and 15R-SiC polytypes. These films exhibit a variety of surface features that include hillocks and depressions. Structural defects that occur include the micropipes and dislocations that propagate from the substrate into the film (Powell et al., 1994). Although excellent devices have been fabricated using these films, much work remains to improve the surface morphology and to reduce the defect density.

The electrical quality achievable in SiC epitaxial CVD films was significantly improved recently by the development of the "site-competition epitaxy" process by Larkin et al. (1993) at the NASA Lewis Center. In this process, the incorporation of nitrogen and aluminum into a SiC epilayer grown on a silicon-face vicinal (0001) plane is controlled by setting the silicon/carbon ratio in the precursor gases to appropriate values. The nitrogen donor atoms that reside on carbon sites in the SiC crystal lattice compete with carbon atoms during growth. Increasing the carbon concentration (i.e., decreasing the silicon/carbon ratio) decreases the nitrogen incorporation in the epilayers. On the other hand, the aluminum acceptor atoms that reside on silicon lattice sites compete with silicon atoms during growth. Increasing the silicon

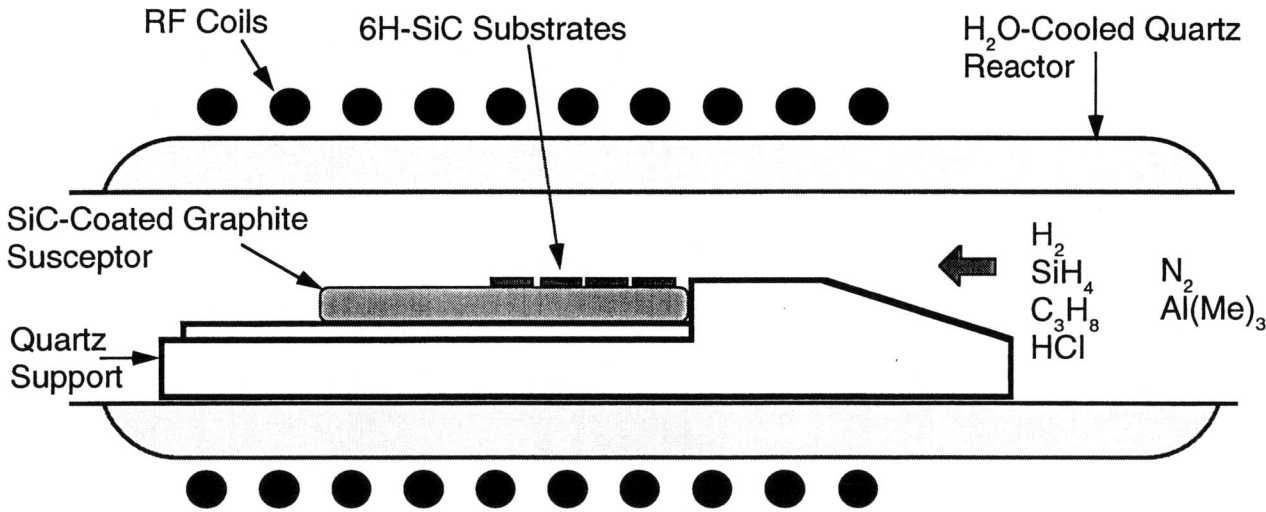

FIGURE 2-7 Schematic of a typical SiC CVD growth chamber. SOURCE: Powell (1993).

concentration (i.e., increasing the silicon/carbon ratio) reduces the aluminum concentration in the epilayers. Undoped films with net carrier concentrations in the low 10^{14} cm^{-3} range can be achieved. In addition, both n-type and p-type epilayers with carrier concentrations in the 10^{14} cm^{-3} range to greater than 10^{19} cm^{-3} can also be consistently achieved. 6H-SiC p-n-junction diodes with record reverse breakdown voltages of greater than 2,000 V were fabricated using this new site-competition epitaxy process (Neudeck et al., 1993).

SiC Epitaxy in the A-axis Direction. In work at Westinghouse, growth directions perpendicular to the c-axis (0001) direction were investigated (Hobgood, 1993). Two important nonequivalent directions, the a-axis (1210) direction and the prismatic (1010) direction, were studied. An advantage of the a-axis growth direction is that the polytype stacking sequence is contained within the growth surface, so homo-epitaxy is always achieved. Another motivation for this work is that the crystal properties vary in different directions, so specific devices require specific crystal orientation for optimum performance. The results of this investigation found that excellent quality films were obtained with growth in the a-axis direction, whereas relatively poor films were obtained in the prismatic direction. Also, it was found that good quality films could be grown at lower temperatures in the a-axis direction compared to growth in the c-axis direction (using off-axis (0001) substrates). In comparing a-axis and c-axis growth, it was found that the growth rates were about the same.

In the a-axis growth, the dopant incorporation was also found to be a function of the silicon/carbon ratio in the precursor gases. The behavior was similar to that for c-axis growth on the silicon-face (0001) plane. However, the doping was shifted toward more n-type doping compared to c-axis growth. Doping levels from a low of 10^{15} cm^{-3} to more than 10^{19} cm^{-3} were achieved for n-type doping. For p-type doping, levels up to 5×10^{18} cm^{-3} were achieved.

Hetero-epitaxial Growth of 3C-SiC Films. Prior to the availability of large-area, high-quality SiC substrates, many growth experiments have been conducted on non-SiC substrates. Early in the 1980s, large-area, single-crystal films of 3C-SiC were achieved by CVD on (001) silicon substrates (Nishino et al., 1983). Unfortunately, the 3C films grown on silicon had a high defect density, which included stacking faults, microtwins, and a fault known as inversion domain boundaries (IDBs), also known as antiphase boundaries. These defects were thought to be caused by the 20 percent lattice mismatch between silicon and SiC or perhaps by the nucleation process. Further work demonstrated that the IDBs could be eliminated by using vicinal (001) silicon substrates with tilt angle in the range 0.5° to 4°. Although the films were now free of IDBs, they still contained a high density of stacking faults and other defects. Devices fabricated from these 3C films have not achieved satisfactory performance.

To eliminate the problem of the large lattice mismatch, titanium carbide (TiC_x) with a lattice match within 1 percent was investigated (Parsons, 1987). Somewhat improved growth of 3C-SiC films was reported, but great difficulties in producing defect-free, single-crystal TiC_x has hindered its use as a substrate for SiC growth.

In a previous section, it was pointed out that 3C-SiC generally grows on vicinal (0001) α-SiC with small tilt angles if there is contamination or defects on the growth surface. Unfortunately, 3C-SiC films grown in this manner typically have a defect known as double-positioning boundaries. This defect arises because there are two possible orientations of the 3C-SiC film that can nucleate on an α-SiC substrate; these two orientations are rotated 180° about the c-axis with respect to each other. When nuclei with both orientations occur on the substrate, the intersection of domains with different orientations are not coherent and they form double-positioning boundaries that are electrically and chemically active.

Recent work at Kyoto University has shown that the density of double-positioning boundaries in 3C-SiC films grown on vicinal (0001) 15R-SiC is less than that found in 3C-SiC films grown on 6H-SiC (Chien et al., 1994). Chien and colleagues presented a model that predicts 3C-SiC films that are tens of micrometers thick and grown on (0001) 15R-SiC should be free of double-positioning boundaries. Unfortunately, the stacking-fault density appears to be very high in these 3C-SiC films.

Another approach investigated at the NASA Lewis Center is to limit the epitaxial growth areas to small mesas on vicinal (0001) 6H-SiC substrates and then limit the nucleation of 3C-SiC to the highest atomic planes on the mesa (Powell et al., 1991). With nucleation limited to a very small region on each mesa, 3C-SiC films will grow laterally and will subsequently cover the mesa with a double-positioning boundary-free 3C-SiC film. This approach has been successful obtaining double-positioning boundary-free 3C-SiC films on 1 mm² mesas. These films also have a lower stacking-fault density than previously reported 3C-SiC films grown on SiC substrates. Combining this technique with the site-competition epitaxy process for doping SiC epitaxial films, p-n-junction diodes with reverse breakdown voltages exceeding 300 V were fabricated (Neudeck et al., 1993). This breakdown voltage is four times that of any previously reported 3C-SiC diode.

Other Epitaxial Processes. The sublimation sandwich process (Ivanov and Chelnokov, 1992) is similar to the modified sublimation process. In the sublimation sandwich process, the substrate is placed near a solid SiC source that is sublimed at temperatures greater than 1800 °C. The resulting vapor condenses on the substrate that is held at a slightly lower temperature. The high temperature required by this process is its main disadvantage.

In the liquid-phase epitaxy technique (Ivanov and Chelnokov, 1992), the substrate is placed in liquid silicon that is saturated with carbon at a temperature in the range of 1500-1700 °C. If the temperature is lowered, SiC is deposited from the supersaturated silicon solution onto the substrate. In one version of this process, the liquid silicon solvent is suspended by an electromagnetic field; this "containerless" approach avoids contamination of the solvent by a crucible. The higher temperature required and the difficulty of control are disadvantages of this approach.

Summary. Excellent epitaxial films of α-SiC polytypes can now be grown on α-SiC substrates. Both n-type and p-type films with net carrier concentrations from 10^{14} cm^{-3} to greater than 10^{19} cm^{-3} can be routinely achieved. The growth of large-area epilayers that are free of micropipes will only be possible when micropipe-free substrates are available. In the future, it will probably be desirable to reduce the growth temperature from the present 1450 °C; this may be beneficial for some device fabrication processes.

NITRIDE MATERIALS

There are four major nitride semiconductors and several minor ones. The four major nitride semiconductors are indium nitride (InN), gallium nitride (GaN), aluminum nitride (AlN), and boron nitride (BN). For high-power electronics applications, there is yet another nitride (iron nitride) that, although not a semiconductor, warrants attention. These materials are composed of cations from Group III of the periodic table and a nitrogen anion from Group V. They are often referred to as III-N materials. AlN, GaAlN, and GaN have been studied for some time, but due to the lack of good single crystals, the electronic, optical, and physical properties of single-crystal nitrides are not extensively known. Interest in the nitride materials has dramatically increased with the recent introduction of bright blue light-emitting diodes (LEDs) by Nichia, the successful growth of better samples, and the accumulation of more precise data (Strite and Morkoc, 1992; Choyke and Linkov, 1993; Lin et al., 1994; Morkoc et al., 1994).

Properties

The most intriguing aspect of the large bandgap nitrides (i.e., AlN, GaN, and InN) is the fact that they form a continuous alloy system with room-temperature direct bandgaps varying from 6.2 eV for AlN to 3.45 for GaN, to 1.9 for InN. In addition, there is a small lattice mismatch (<1 percent) between (wurtzite, 2H) AlN and 2H-SiC, and between cubic BN (cBN) and diamond. The band structure of the hexagonal and cubic modifications of AlN and GaN are given in Figures 2-8 and 2-9 (Lambrecht and Segall, 1992).

Boron nitride is the least understood of the nitrides. Most work is directed towards the synthesis and characterization of cBN as it is believed to exhibit an indirect bandgap in excess of 6.4 eV. The relative dielectric constant of cBN is 6.5 and its hardness is 4,500 kg/mm^2 compared with 3,980 for SiC and 10,400 for diamond (Davis, 1992). Its thermal conductivity is believed to be 1,300 W/m·°C, or more than twice that of SiC and about 60 percent that of diamond. Young's modulus is 5.2 MPa compared with 4.0 MPa for SiC. The Poisson ratio for cBN is 0.2 or equal to that of both diamond and SiC. The thermal expansion coefficient of cBN is 3.7×10^{-4}/°C, which is the same as that of SiC but greater than the 2.3×10^{-4} of diamond. Unfortunately, cBN has also been the most difficult to synthesize.

Aluminum nitride exhibits a direct bandgap of 6.2 eV in its hexagonal form. Like diamond, AlN exhibits negative electron affinity (Benjamin et al., 1994). While cubic AlN has recently been synthesized as a thin film on cubic (3C) SiC, its bandgap has not been ascertained but is believed to be somewhat less than 6.2 eV and is most probably indirect. The thermal conductivity of polycrystalline AlN is 3.0 W/cm·°C at room temperature—over twice that of silicon and 60 percent that of SiC. Its relative dielectric constant is 10.0, or 85 percent that of SiC. The best crystallinity reported to date using X-ray $\theta/2\text{-}\theta$ diffraction data is 90 arc-seconds for films grown hetero-epitaxially on sapphire. While there are reports in the literature of both n- and p-type AlN having been synthesized, these reports are not recent (Chu et al., 1967; Rutz, 1976). Aluminum in AlN has an affinity for oxygen and oxygen appears at a deep level in AlN. Oxygen is typically found in AlN in concentrations of 10^{20} cm^{-3}, rendering it extremely difficult to obtain AlN with either p- or n-type conductivity.

There is currently a considerable amount of work underway addressing the AlN doping issue. Alloys of AlN and SiC have recently been made. These may not be true alloys, however, as there is no measurable interdiffusion at temperatures up to 1900 °C. Nevertheless, absorption-band edge measurements on this alloy appear to track the mole-fraction composition. The low mass of nitrogen engenders AlN with a high optical phonon energy; for this reason, the charge-carrier velocity could be very high and approach that of diamond.

The lattice parameters for AlN are a = 3.112 Å and c = 4.982 Å (293 K). The AlN linear thermal expansion coefficient is $a_\perp = 5.27 \times 10^{-6}$ K^{-1}; T = 20-800 °C; and $a_\parallel = 4.15 \times 10^{-6}$ K^{-1}. The thermal conductivity of AlN is k = 2 W/cm·°C at room temperature. The density of AlN is d = 3.244 g/cm^3. Phonons for AlN are in the frequency range of 895 cm^{-1} to 303 cm^{-1}. The dielectric constants for AlN are $\epsilon(0) = 9.14$ and $\epsilon(\infty) = 4.84$ (300 K).

Gallium nitride exhibits a direct bandgap of 3.5 eV in its hexagonal form and apparently slightly less in its cubic form. Its lattice constant is 3.189 Å, or about 4 percent greater than that of SiC. Its dielectric constant has been measured at 8.9 to 9.5, or just less than that of SiC. It is thought to have an effective electron mass of 0.20, but this figure should not yet be taken as definitive. Along the a-axis, the coefficient of thermal expansion is 5.5×10^{-6} K^{-1}. The thermal conductivity at room temperature of GaN is 1.3 W/cm·°C, nearly equal to the 1.45 W/cm·°C of silicon and about three times higher than GaAs. After two decades of research, both p-type and n-type GaN have now been produced in hetero-epitaxial thin films. In the best of this material (e.g., with X-ray $\theta/2\text{-}\theta$ diffraction data exhibiting a full width at half maximum of about 27-28 arc-seconds; Plano et al., 1994), acceptors freeze out at about 205 K and holes exhibit a mobility of 450 cm^2/V·s. Electron mobilities of 1,200 cm^2/V·s at room temperature have been observed. Both of these values exceed those of SiC but not of diamond. With better crystallinity, these mobilities may perhaps be

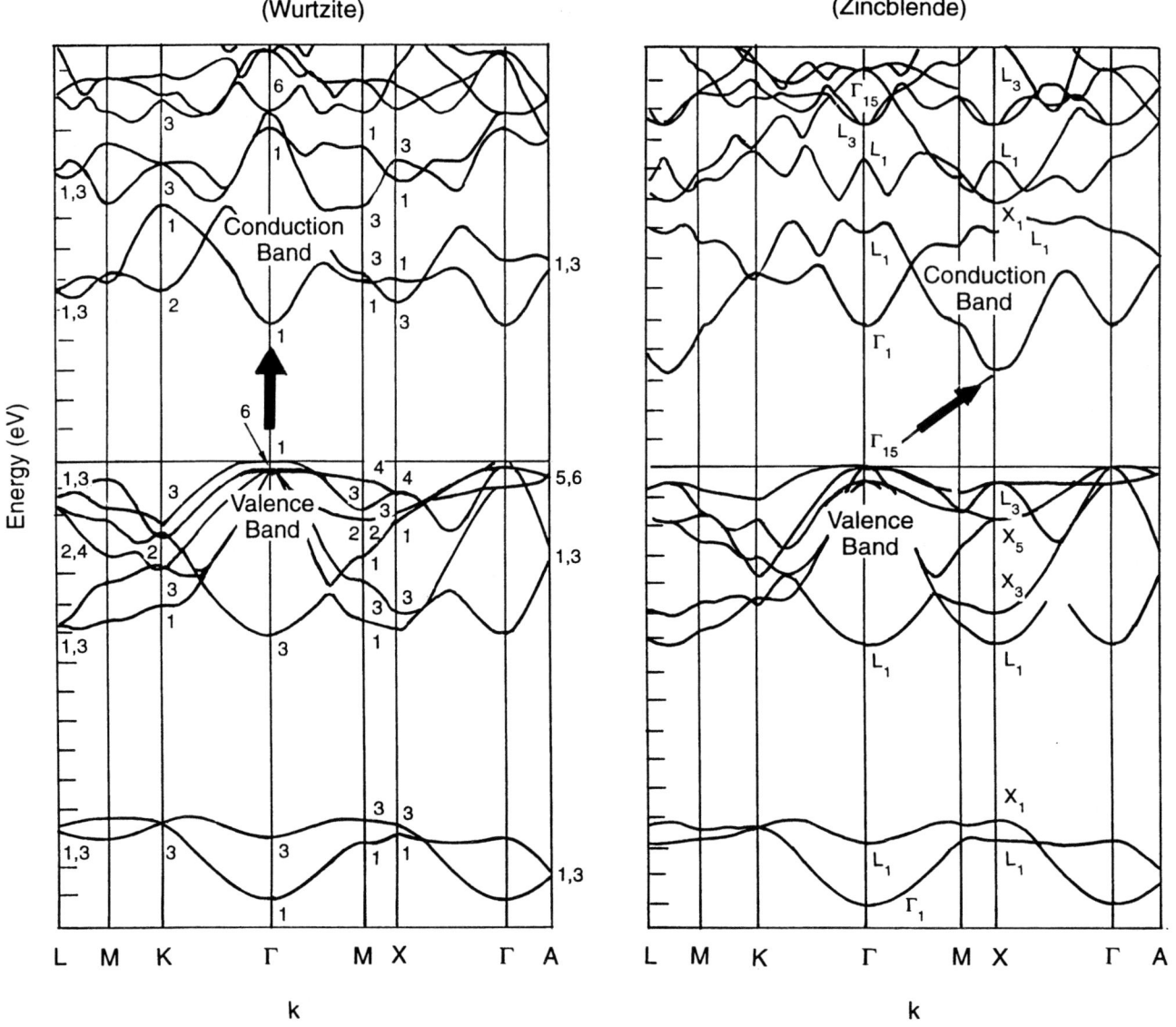

FIGURE 2-8 Band structure of hexagonal and cubic modifications of AlN. SOURCE: Based on Lambrecht and Segall (1992).

expected to further improve. The dielectric strength is believed to be about equal to that of SiC and its computed peak electron velocity exceeds 2×10^7 cm/s.

A large number of luminescent features have been reported between 1.65 and 3.5 eV in GaN. These have been attributed to a variety of impurities and defects. However, there is a great deal of controversy in the literature as to the various interpretations. A number of articles have reviewed the literature of luminescence and absorption lines in AlN and GaN (Strite and Morkoc, 1992; Choyke and Linkov, 1993). Although there is currently great commercial potential for GaN optical devices, the beneficial impurities and defects of this material for luminescent features are detrimental for high-temperature device operation.

Indium nitride exhibits a direct bandgap of 1.9 eV and an indirect bandgap only slightly higher. Its thermal conductivity and most other properties have not yet been definitively ascertained. Its lattice constant of 3.5 Å considerably exceeds that of SiC, AlN, and GaN. Double heterostructures of GaN/InGaN/GaN currently exhibit the brightest purple, blue, and blue-green LEDs ever made. The blue and blue-green devices are commercially available from Japan and exhibit an operating efficiency of about 2.7 percent.

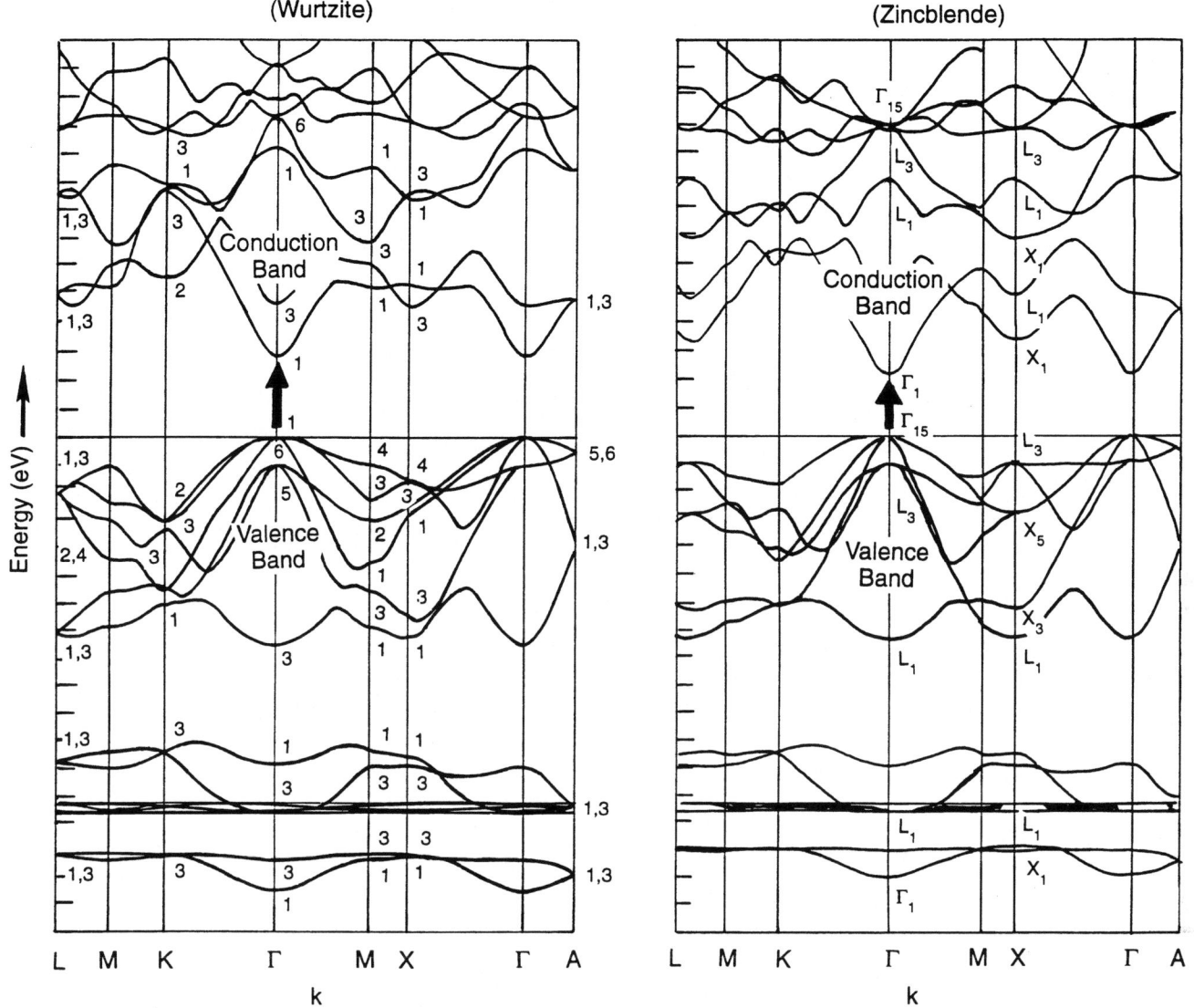

FIGURE 2-9 Band structure of hexagonal and cubic modifications of GaN. SOURCE: Based on Lambrecht and Segall (1992).

The nitride materials, like diamond, are very difficult to etch with liquid etchants. This is an active area of research, however, and phosphoric acid and sodium hydroxide have recently been found to work on both AlN and GaN. Also like diamond, the nitride semiconductors can be left exposed to the atmosphere at high humidity for months at a time without becoming oxidized or otherwise having their surface properties changed. Unlike GaAs, the nitrides do not exhibit self-depleting surface states. For this reason, devices employing p-n junctions do not exhibit high surface recombination velocities, which should lead to a long laser-operating life and to extremely long charge-storage capability (e.g., millennia) and extremely low-leakage devices suitable for applications such as nonvolatile memories.

Crystal Growth

Very little work has been done in attempting to synthesize boules of the nitrides. Japanese researchers have synthesized small boules of cBN via high-temperature, high-pressure processes and have even made light-emitting p-n junctions of the material. The material was so contaminated by impurities, however, that the absorption band edge was not very sharp and its bandgap was difficult to ascertain. The p-n junctions emitted light in both the visible and in the ultraviolet. Most of the cBN research in the United States has been directed at thin-film synthesis on diamond. It is exceedingly difficult to obtain films thicker than 20 nm that are not polycrystalline or fractured. The only high-temperature, high-pressure attempts to synthesize boules of GaN have been in Poland (Perlin, 1993)—however, the boules were very small. Some boule growth has been attempted with AlN, but these efforts have generally not been successful to date.

DIAMOND

Materials Description and Properties

Diamond has been admired as a jewel since antiquity and has been studied for a very long time. In fact, Sir Isaac Newton made measurements of the index of refraction of diamond some time around 1665. Large single crystals are found in nature, and synthetic diamonds have been made in high-temperature, high-pressure anvil machines for about 40 years. However, from a semiconductor standpoint, only limited impurity and defect control has been possible to date. Low-temperature and low-pressure polycrystalline film growth has been actively pursued in the last 10 years, but no high-quality, single-crystal films have yet been obtained. The major properties of crystalline diamond are well understood, and two excellent books chronicle the development of artifact diamond and tabulate its known properties (Davis, 1992; Spear and Dismukes, 1994).

Diamond is an indirect-gap semiconductor, with the lowest minima of the conduction band being located along the delta axes (k = 0.76[1,0,0]). The valence band maximum has a structure that is common to all Group IV semiconductors. There are three bands that are degenerate at the Γ point when spin is neglected. A band calculation by Chelikowsky and Louie (1984) is shown in Figure 2-10. The indirect energy gap at room temperature is 5.5 eV, and between 135 K and 300 K the variation of the bandgap with temperature is given by -5×10^{-5} eV/K. The exciton binding energy, E_x, is about 80 meV.

Effective masses have been measured and calculated in high-quality bulk diamond crystals. The electron effective masses are $m_\perp = 0.36\ m_o$, and $m_\parallel = 1.4\ m_o$. The effective masses of the holes are $m_h = 1.08\ m_o$, $m_l = 0.36\ m_o$, and $m_{so} = 0.15\ m_o$. Hall mobilities have been obtained for n- and p-type diamond as follows: $\mu_e = 2{,}200$ cm^2/V·s (RT), and $\mu_h = 1{,}600$ cm^2/V·s (RT).

Diamond is famous for its excellent thermal conductivity. In the last few years single isotope diamond has been produced, and it has a higher thermal conductivity than natural diamond (i.e., 32 W/cm•K; Anthony, 1994). Ordinary isotopic ratio diamond has a thermal conductivity as shown in Figure 2-11. The dielectric constant of diamond measured at 300 K is 5.5. The lattice parameter a is 3.56683 Å at 298 K. Linear thermal expansion coefficients for various temperatures are

$\sim 1 \times 10^{-6}$ K^{-1} (300 K),
$\sim 3 \times 10^{-6}$ K^{-1} (600 K),
$\sim 4 \times 10^{-6}$ K^{-1} (900 K), and
$\sim 5 \times 10^{-6}$ K^{-1} (1200 K).

The density of diamond is 3.51525 g/cm^3, as calculated from the lattice constant. The second-order elastic constants for diamond are $c_{11} = 10.764 \times 10^{12}$ dyne/cm^2 (296 K), $c_{12} = 1.252 \times 10^{12}$ dyne/cm^2 (296 K), and $c_{44} = 5.774 \times 10^{12}$ dyne/cm^2 (296 K).

Methods of Synthesis and Characterization

Synthesis

Aside from the high-pressure, high-temperature synthesized boules of diamond, virtually all diamond films are grown by plasma-assisted methods in the presence of an abundance of atomic hydrogen. The feedstock typically consists of 99 percent H$_2$ and 1 percent CH$_4$. There are many variations of this basic method. While conventional (lower bandgap) materials are typically synthesized at substrate temperatures approaching two-thirds of the melting temperature, this is not possible with diamond and some of the higher bandgap materials. The typical substrate temperature of 900 °C cannot be used to assure that feedstock species alighting on the growth surface are fully "activated." The activation-energy threshold (E_a) is generally composed of three parts: (1) energy sufficient to dissociate the molecule of radical, (2) energy sufficient to chemisorb rather than physisorb the feedstock species, and (3) energy sufficient to ensure that the adsorbed species arrive at a proper lattice site. In diamond-film synthesis,

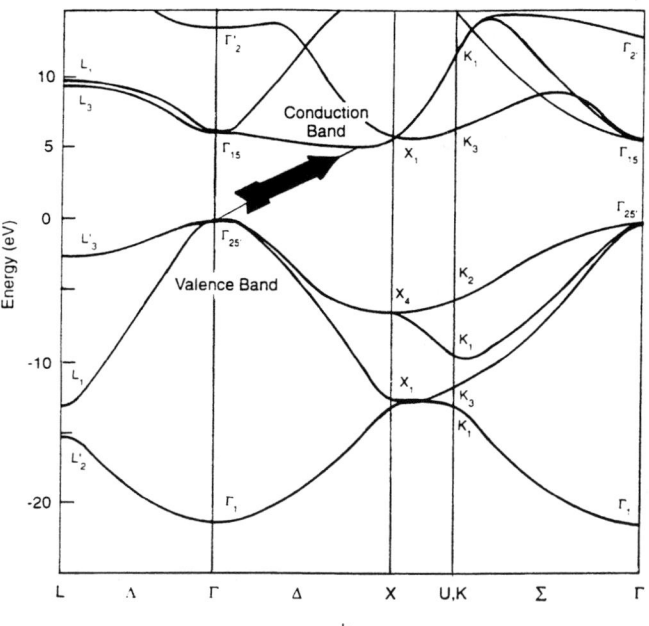

FIGURE 2-10 Band-structure calculation of diamond.
SOURCE: Based on Berman and Martinez (1976).

The cost and safety aspects of methane and gaseous hydrogen storage are in some cases circumvented by an alcohol and water plasma process. Several halogen-based techniques are also currently under investigation to reduce the synthesis cost. The target for diamond is to grow large areas of single crystals. This has not yet happened. Single crystalline diamond has not yet been grown on any substrate except natural diamond and cBN. Cubic BN is much less plentiful than natural diamond and available only in very small sizes. Attempts to synthesize diamond on all other foreign substrates has resulted in films characterized by numerous low-angle grain boundaries and much worse. Large area arrays of seeded natural diamond on silicon has resulted in mosaic diamond films.

Characterization

The preferred method of characterization of the diamond films is by Raman spectroscopy. An unstrained diamond film exhibits a Raman signal at 1,332 cm^{-1}. The full width at half maximum of this Raman signal is an

this is typically accomplished by providing kinetic energy to the feedstock species and to the atomic hydrogen. This kinetic energy is typically imparted via a plasma. These plasmas are generated by direct current (including arc jets), radio frequency fields, microwave fields, or by combustion (e.g., oxy-acetylene torch). Diamond-film synthesis has an additional complication unknown to the synthesis of any other semiconductor. The lowest energy form of its surface is that of an sp$_2$-configured π bond that is graphitic in character. When the diamond surface is so constructed, only graphite can be grown on it. To prevent this unwanted surface construction, the surface is terminated by hydrogen, but this hydrogen must be removed and quickly replaced by carbon to grow the diamond. Since hydrogen bonds to the diamond surface with an energy of 104 kcal/mol versus the 88 kcal/mol of the carbon-carbon bond, it is not easy to remove. Removal requires energetic hydrogen in a "sea" of hydrogen atoms and (typically) methyl radicals or acetylene. For every 10^4 hydrogen atoms removed from the diamond surface, only one is replaced by a carbonaceous radical; the remainder are replaced by another hydrogen. The growth process is thus slow and relatively inefficient, although DC arc jets and combustion jets have grown diamond at rates exceeding 100 micron/h.

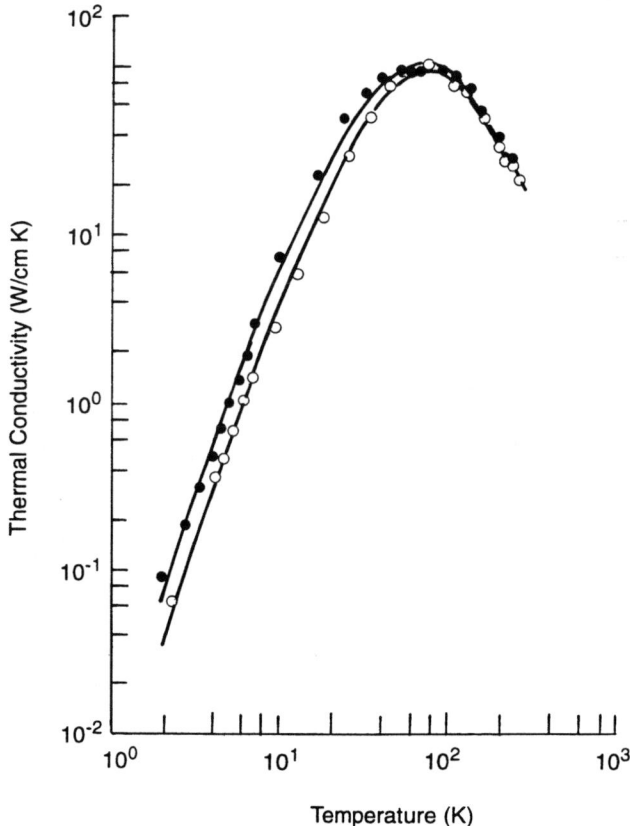

FIGURE 2-11 Thermal conductivity of two Type IIa diamonds.
SOURCE: Based on Berman and Martinez (1976).

indication of the quality of the film. The best of type IIa natural diamond (diamond with no active optical centers) typically exhibits a full width at half maximum of 2.2 cm^{-1} or slightly less. The best of artifact homo-epitaxial diamond films was grown by microwave plasma-enhanced CVD and was characterized by a Raman signal of 1.7 cm^{-1}. The full width at half maximum Raman signature has been correlated with the thermal conductivity of diamond. In polycrystalline diamond films, a full width at half maximum Raman signature of 3.2 cm^{-1} generally ensures that the thermal conductivity (in the direction of growth) exceeds 17 W/cm·K and the electrical resistivity exceeds 10^6 Ω·cm. Diamond films can be grown with intrinsic resistivities of 10^{10} Ω·cm and larger, but at slower growth rates (e.g., < 1 µm/h).

Diamond Processing

Diamond is not etched by boiling acids or bases. There are two preferred methods of processing diamond. The first is the use of kinetic energy beams of oxygen or oxygen-containing molecules or radicals. The second method is by electrolytic etching. The electrolytic etching is generally limited to removal of defect-ridden or otherwise conducting regions of diamond. Boron is the only universally recognized acceptor impurity that can be controllably introduced into diamond. Until recently, only a small portion of the boron in diamond was electrically activated. It can now be introduced and nearly 100 percent electrically activated by a series of implantation processes to concentrations of 1 x 10^{19}/cm^3. By a similar procedure the same investigator stated that he has activated phosphorous in diamond at 80 MeV (Prinz, 1994).

3

Device Physics: Behavior at Elevated Temperatures

HIGH-TEMPERATURE EFFECTS: FUNDAMENTAL, MATERIALS-RELATED PROPERTIES

Although technological considerations such as gate and ohmic contact metallization, as well as dopant profiles are an inextricable component of device behavior at elevated temperatures, it is useful to specifically focus on the device physics expected at elevated temperatures, aside from the technologically relevant, but perhaps addressable issues, such as increased diffusion of dopants. The fundamental device physics, dependent on the materials properties, include parameters such as carrier concentrations, scattering, and leakage current, and are described below.[1]

Carrier Mobilities[2]

The mobilities of the charged carriers will determine the speed of operation of the devices, and those mobilities are expected to change with increased temperature of operation. There are a variety of carrier scattering mechanisms in the semiconductor material that will limit carrier mobilities: principal among these are acoustic phonon scattering (also referred to as lattice scattering and ionized impurity scattering), which refers to the Coulombic interactions between the charged carriers (electrons or holes) with the ionized dopants that give rise to the free carriers. To a first approximation, for a given material, the mobility limitation from ionized impurity scattering scales with temperature as $T^{3/2}$ (that is, the mobility decreases at lower temperatures), while that from acoustic phonon scattering scales as $T^{-3/2}$ (mobility decreases at higher temperatures). However, the mobility also depends on factors such as the density of ionized impurities, which also varies with temperature. In addition, for polar semiconductors, such as GaAs, optical phonon scattering becomes important. Moreover, the critical, current-carrying area of some devices comprises a thin sheet of charge near a hetero-interface. This characterizes the inversion-layer charge at the interface of silicon and silicon dioxide in metal-oxide semiconductor field effect transistors (MOSFETs). In this case, other factors will affect the charge scattering and hence the mobility. Some of these factors include the presence of charged defects in the oxide, or at the interface, as well as morphological roughness of the interface. The total mobility at a given temperature is then determined by summing up all the significant contributions to carrier scattering.

Figure 3.1 shows the calculated dependence of mobility on temperature for undoped 6H-SiC and 3C-SiC (Shur et al., 1993). The mobility of the 6H-SiC decreases from 420 cm^2/V·s at 25 °C to around 120 cm^2/V·s at 200 °C. The room-temperature mobility for SiC doped in the 10^{17} cm^{-3} range is expected to be lower (~250 cm^2/V·s), since the impurity scattering has been increased. Figure 3.2 shows the calculated electron mobility as a function temperature for n-type GaN, doped 10^{17} cm^{-3}.

Intrinsic Carrier Concentrations: Dependence on Bandgap Energy and Temperature

Thermal energy, kT, can be sufficient to promote electrons from the valence band to the conduction band, giving rise to a thermally generated current, which is not

---

[1] For a more detailed consideration of some of the properties described in this chapter, it is suggested that a basic semiconductor text be consulted, such as S.M. Sze's *Physics of Semiconductor Devices* (1981).
[2] Further discussion on this topic can be found in Ridley (1993).

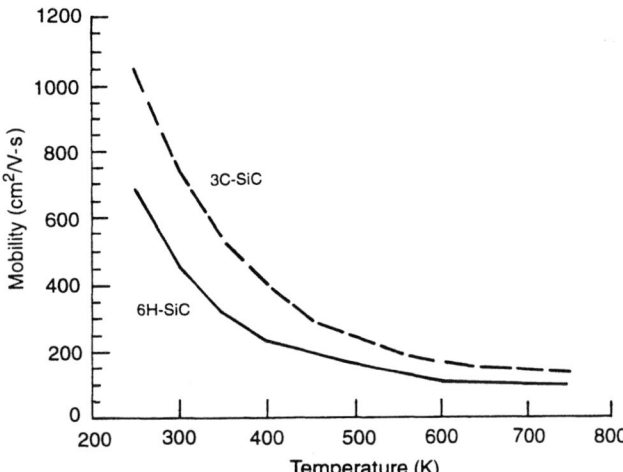

FIGURE 3-1 Calculated electron mobility as a function of temperature for undoped 6H-SiC and 3C-SiC. SOURCE: Shur et al. (1993).

controlled by explicit device operation. In fact, at sufficiently high temperatures, the thermally generated intrinsic carrier concentration can exceed the density of explicitly introduced dopants: thus the ability is lost to control the concentration of the charge carriers in the device. The intrinsic carrier density for a given material is given by

$$n_i = \left(2\pi \frac{kT}{h^2}\right)^{3/2} (m_{dh} m_{de})^{3/4} \exp\left(-\frac{E_g}{2kT}\right), \quad (3.1)$$

where k and h are Boltzmann's and Planck's constants; T is the absolute temperature; m_{de} and m_{dh} are the electron and hole effective masses, expressed in multiples of the free electron mass; and E_g is the bandgap energy. In an intrinsic semiconductor, $n_i = n = p$, where n is the electron density and p is the hole density. The simple effect of increased temperature on Equation (3.1) is to increase n_i, hence leakage current. This is a critical concern for technologies that depend on an insulating or semi-insulating substrate, which is the case for GaAs metal semiconductor field effect transistors (MESFETs). A semi-insulating GaAs substrate will show approximately six orders-of-magnitude decrease in bulk resistivity from 25-300 °C, resulting in a substrate leakage current that cannot be controlled by the device gate (Look, 1989; Lee et al., 1995).

For a given temperature, the intrinsic carrier density is less for a larger bandgap material. These data are plotted in Figure 3.3 for silicon, GaAs, and SiC (the temperature dependence of E_g is neglected here). As the temperature is increased, n_i increases rapidly. Moreover, most semiconductors have bandgaps whose magnitudes decrease as the temperature is increased. This is shown in Figure 3.4 for silicon, doped with either n-type or p-type at different concentrations. For example, the bandgap energy in intrinsic silicon decreases by 0.083 eV from room temperature to 300 °C. This represents an increase in intrinsic carrier density at 300 °C of 2.5 times over the room-temperature intrinsic carrier density.

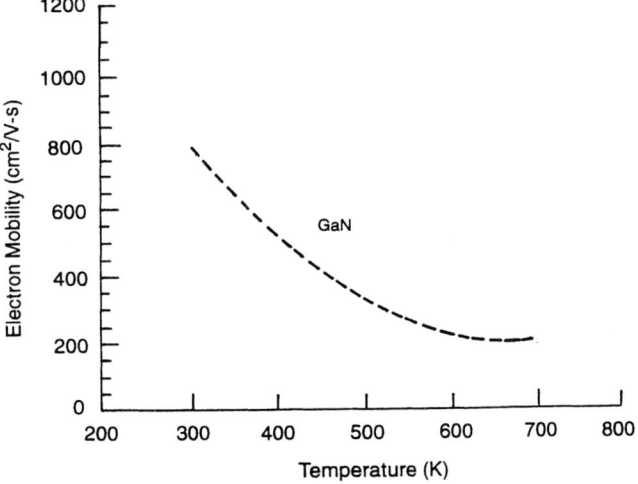

FIGURE 3-2 Calculated electron mobility as a function of temperature for GaN doped n-type, 10^{17} cm^{-3}. SOURCE: Shur et al. (1993).

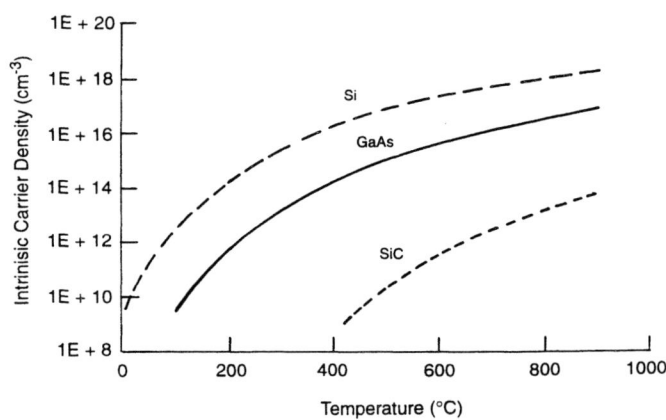

FIGURE 3-3 Intrinsic carrier density for silicon, GaAs, and SiC.

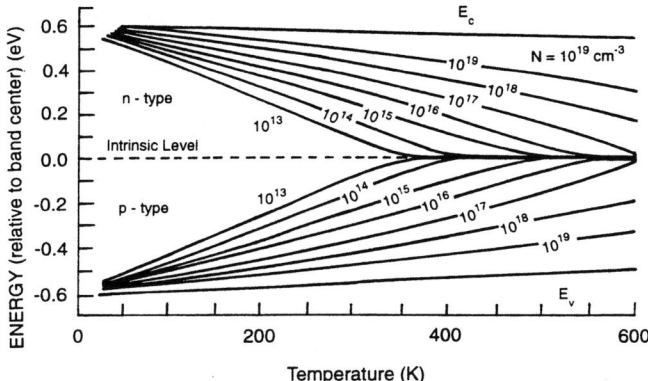

FIGURE 3-4 Decrease in silicon bandgap with increasing temperature. SOURCE: Sze (1981).

PREDICTING HIGH-TEMPERATURE-DEVICE PERFORMANCE: MATERIALS-RELATED FIGURES OF MERIT

The high-temperature effects described above have been reported to have first-order effects on device operating characteristics (Khan et al., 1993b), which are described in greater detail below. Figures of merit, based on the materials parameters affecting devices, provide a useful basis of comparison for the possible technologies. In the figures of merit discussed below, the highest values possible are desirable. Johnson's Figure of Merit (JFM; Johnson, 1965) relates to the frequency and power product of a semiconductor transistor, and is given by

$$JFM = \left(\frac{E_m v_s}{2\pi}\right)^2, \quad (3.2)$$

where E_m is the avalanche breakdown electric field and v_s is the carrier saturated velocity. JFM can also be viewed as the square of the quotient of the breakdown voltage of a semiconductor layer and the intrinsic transit time for carriers moving through the layer. JFM accounts for the fact that in an intrinsic device (i.e., one without parasitic resistance or reactance) there is a tradeoff between the time a carrier spends gaining energy in an electric field as it drifts through a device and the response time of the device. JFM is related to electronic properties and does not account for thermal effects.

The Keyes' Figure of Merit (KFM; Keyes, 1972) takes into account the thermal properties of a material and is given by

$$KFM = \lambda \left(\frac{c v_s}{2\pi \epsilon}\right)^{1/2}, \quad (3.3)$$

where λ is the material thermal conductivity, c is the velocity of light in vacuum, and ϵ is the material dielectric constant. Keyes assumes that smaller devices are inherently faster in response at fixed electronic-input impedance level. However, devices cannot be made smaller without increasing the thermal resistances, thereby limiting the power output and introducing thermal conductivity as a factor. The breakdown field is not significant in this figure of merit since KFM addresses a thermal rather than an electronic limit.

Baliga (1982) noted the role of the saturation velocity, v_s, in both the JFM and KFM, which is important for high-speed electronics but not necessarily the major parameter for devices to be used in power applications. Instead, he emphasized the role of high carrier mobility and large electric field at breakdown (E_m). For lower-frequency power devices, where conduction loss in the on-state is the dominant power loss, the figure of merit is

$$BFM = e\mu E_m^3. \quad (3.4)$$

At higher frequencies, switching losses due to charging and discharging of the device capacitance assumes greater importance. In that case,

$$BHFM = e\mu E_m^2. \quad (3.5)$$

Recently, Chow and Tyagi (1994) have carried out a comparison of figures of merit of various semiconductors for high-power and high-frequency unipolar devices. A portion of their results are shown in Table 3-1. The calculations are made at room temperature, rather than at some elevated temperature of operation, as has been the norm for these calculations. It is generally a safe assumption that the materials advantages leading to higher figures of merit at room temperature will persist at higher temperatures. Breakdown voltages, mobilities, and saturation velocities will be degraded in all cases, but should fall off with temperature less precipitously for the wide bandgap materials than for silicon or GaAs. For JFM, the high breakdown field dominates, making all of the wide bandgap materials attractive compared to silicon, germanium, and GaAs. Since the figures of merit weight various properties, they do not appear to provide a clear

TABLE 3-1 Comparison of Normalized Figures of Merit of Various Semiconductors for High-Power and High-Frequency Unipolar Devices

Material	JFM	KFM	BFM	BHFM
Silicon	1.00	1.00	1.00	1.00
Germanium	—	—	0.13	0.28
Diamond (a)	5,330	31	14,860	1,080
Diamond (p)	6,220	32	11,700	850
AlN ($\mu = 14$)	5,120	2.6	390	14
AlN ($\mu = 1,090$)	5,120	2.6	31,670	1,100
AlAs	630	—	7.3	2.0
GaAs	11	0.45	28	16
GaN	790	1.8	910	100
6H-SiC	260	5.1	90	13
3C-SiC	110	5.8	40	12
4H-SiC	410	5.1	290	34

SOURCE: Chow and Tyagi (1994).

choice among the different wide bandgap materials. The high thermal conductivities of the wide bandgap materials increase the values of KFM. Also, their lower dielectric constants reduce the capacitance per unit area, thereby further increasing KFM.

The figures of merit suggest that devices whose limitations are principally due to their electronic limitations, such as saturated velocity or breakdown electric field, can achieve higher power density in the wide bandgap materials. Field effect transistors are among this class of devices. For devices whose operation is limited by thermal considerations, such as the material thermal resistance, higher power density should also be achievable in the wide bandgap materials. Among these devices are bipolar transistors. These predictions must be moderated by the fact that figures of merit generally provide a rough estimate of performance, since only the "intrinsic" device is considered. The figures of merit do not fully account for parasitic resistance and other detailed effects that limit device performance and that require more careful examination for a particular device technology.

Device Physics at High Temperatures

More-detailed discussions of the effects of high temperature on device performance are given in Appendices A and B for silicon- and GaAs-based technologies operating at high temperatures. Appendix C presents detailed considerations of the influence and issues that surround insertion of the wide bandgap semiconductors into microwave devices. Specific issues of consequence for high-temperature device operation are (1) the effect on device conductivity in the on-state, and (2) the effect of leakage currents in the off-state. The examples below provide an introduction to the issues that are of concern for device technologies, building upon the high-temperature changes in carrier density and mobility.

Junction Leakage: p-n Junctions and Diodes

Shenai et al. (1989) have approximated junction reverse leakage currents to have the following dependence, at elevated temperatures:

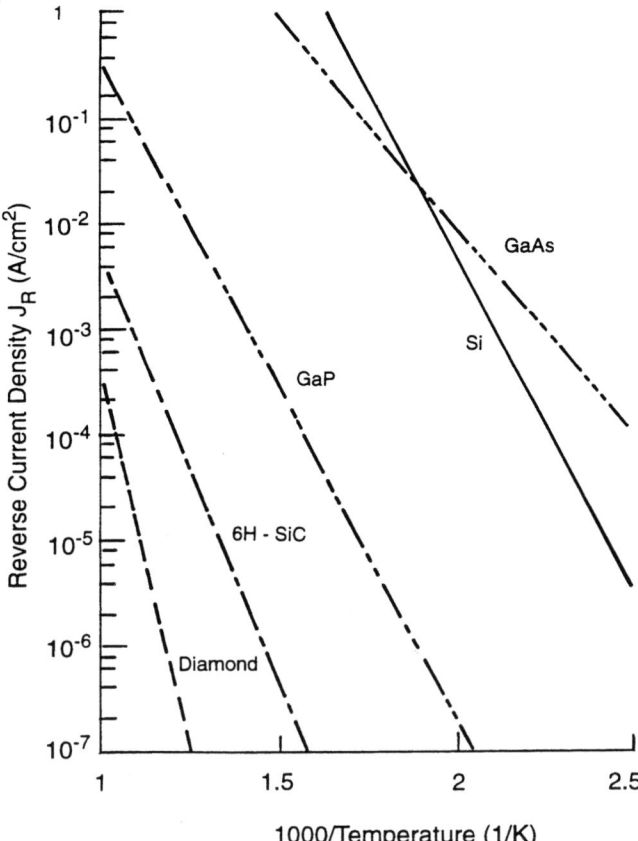

FIGURE 3-5 Calculated reverse leakage current densities in p-n junctions of various materials. SOURCE: Shenai et al. (1989), © 1989 IEEE.

$$J_R \sim J_{G0}\left(\frac{T}{T_0}\right)^{3/2} \exp\left(-\frac{E_g}{kT} + \frac{E_g}{kT_0}\right), \quad (3.6)$$

where J_{G0} is the generation leakage current density at $T = T_0 = 300$ K. The generation current occurs due to thermal generation of electron-hole pairs in the depletion region; the holes and electrons are forced out of the region by the electric field across the region. For the wide bandgap semiconductors, at elevated temperatures, the principal contribution to leakage current is this generation current, rather than diffusion currents that are generally the dominant component for silicon. Hence the diffusion current density has been assumed to be negligible; in the plot of calculated reverse leakage current densities, J_R, shown in Figure 3-5, the results for silicon are only valid for $T > 125$ °C.

Such junction leakage will also affect bipolar devices; collector-base leakage currents increase with increasing temperature. Decreased mobility that arises from increased temperature can cause increased base and collector resistances. This in turn can bring about an increased collector-emitter saturation voltage, $V_{ce(sat)}$, which can easily interrupt circuit functions using bipolar technology (Beasom and Patterson, 1982).

Schottky Leakage

In devices that utilize Schottky barrier junctions, such as MESFETs, reverse leakage currents also increase at elevated temperatures and pose a limitation to device operation. The reverse leakage current of an ideal Schottky barrier diode is approximately given by

$$J_s = A^* T^2 \exp\left(-\frac{q\rho_B}{kT}\right), \quad (3.7)$$

where A^* is an effective Richardson constant and ρ_B is the Schottky barrier height for thermionic emission. For devices, such as the MESFET, where the Schottky junction serves as the gate electrode, it is desired that carrier injection across the barrier under forward bias be minimized, as well as the reverse leakage current. Such forward biased injection is increased at elevated temperatures in both Schottky junctions and p-n junctions.

Threshold Voltage Shifts

The intricate, coordinated interactions of devices within a circuit depend on the control of the threshold voltages of the component devices. Fermi-level changes with temperature reduce the magnitude of the threshold voltage, V_t, as temperature increases. For example, for MOSFETs, the threshold voltage is given by

$$V_T = 2\phi_F - \frac{Q_{ss}}{C_{ox}} + \frac{\sqrt{2\epsilon q N(2\phi_F)}}{C_{ox}} + \phi_{MS}, \quad (3.8)$$

where Φ_{MS} is the metal semiconductor work-function difference, Q_{ss} is the fixed charge located at the silicon-SiO^2 interface, C_{ox} is the gate oxide capacitance per unit area, N is the substrate doping density, and ϵ is the dielectric constant of silicon. Φ_F and Φ_{MS} are temperature-dependent and Q_{ss} has been shown to be temperature independent, at least up to 200 °C. The change in threshold voltage versus temperature for n- and p-channel MOSFET devices is shown in Figure 3-6.

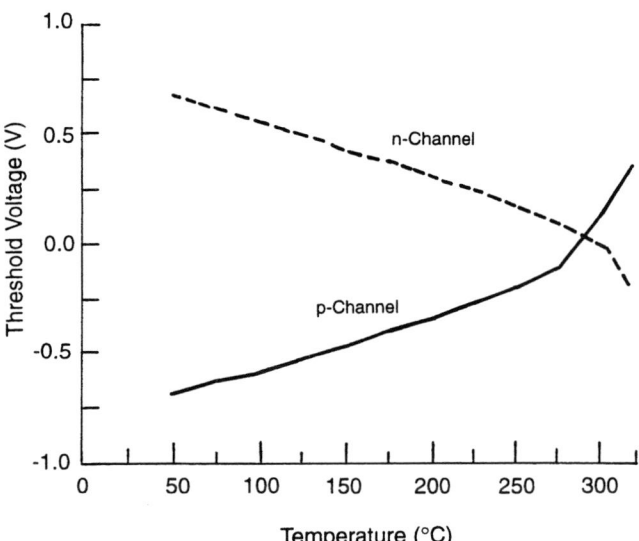

FIGURE 3-6 Variation in threshold voltage versus temperature for n- and p-channel MOSFET devices. SOURCE: D.M. Brown et al. (1994).

Leakage currents represent the ultimate degradation factor with respect to high-temperature operation of silicon MOSFETs and metal-oxide semiconductor (MOS) integrated circuits. At high enough temperatures, the drain to body leakage currents can increase by orders of magnitude and become comparable to the drain channel currents; the transistor can no longer be turned off by the gate. Such leakage lowers noise margins, making the design of static digital devices difficult, and making dynamic circuits and memories impossible to produce. In addition, mobility degradation in MOSFETs at elevated temperatures causes the transistor transconductance to decrease.

Choice of High-Temperature-Device Technologies

The impetus for high-temperature, high-power operation is sufficiently compelling, and there have already been demonstrations of such operation in silicon and GaAs devices. In addition, temperature-dependent extrapolations of device performance and materials-dependent figures of merit allow us to make some assessment of device performance at elevated temperatures for the newer wide bandgap materials such as SiC and GaN. These factors are the basis for the predictions of Figure 3-7, suggesting the materials technologies that will be appropriate for the various temperatures of operation.

In general, the limitations of leakage currents at elevated temperatures places a greater constraint on small-signal devices, compared to digital logic. Reduced mobility and leakage sets a more stringent limitation yet on the performance of microwave devices. Dynamic random access memories (DRAMs) may be further limited by leakage of charge stored in capacitors. We note that the temperature range of -55 °C to 125 °C are the current military specifications that many device technologies must satisfy.

For silicon technology, transistor operation has been demonstrated up to 450 °C (Migitaka and Kurachi, 1994); since leakage currents are already large at this temperature, it is unlikely that silicon devices will be operated at significantly higher temperatures. Since analog-device operation is more strongly affected by the changes in leakage current, gain, and threshold voltages that occur with temperature, it is estimated by the committee that high-temperature operation of these devices will hold good to only 350 °C. Microwave devices, operating at high power densities, would be subject to more severe temperature limitations than small-signal analog devices; hence the committee extrapolates a maximum operating temperature of 200 °C. Power devices have been demonstrated, operating at temperatures as high as 250 °C. Since conventional DRAMs are strongly affected by the leakage currents associated with higher temperature, the committee estimates the maximum temperature of operation to be 150 °C. Devices that were specifically designed for high-temperature operation, with larger charge storage capacitors or based on other memory-cell architectures, perhaps with larger areas, would be capable of a somewhat higher-temperature operation. Finally, reasonable reliability has been demonstrated at 250 °C for bipolar logic (Migitaka and Kurachi, 1994) and at 200 °C for complementary metal-oxide semiconductor (CMOS) logic (Foyt, 1994).

The higher bandgap of GaAs is expected to lead to a somewhat higher-temperature operation than silicon. Transistor operation has indeed been demonstrated at 500 °C (Shenoy et al., 1994). The alloy system AlGaAs, with still larger bandgaps, should allow still higher temperature operation.

Small-signal analog devices have been shown to operate at 300 °C (Bottner et al., 1991); the extrapolation to operation at 400 °C is based on the observation that analog applications are more strongly affected than digital

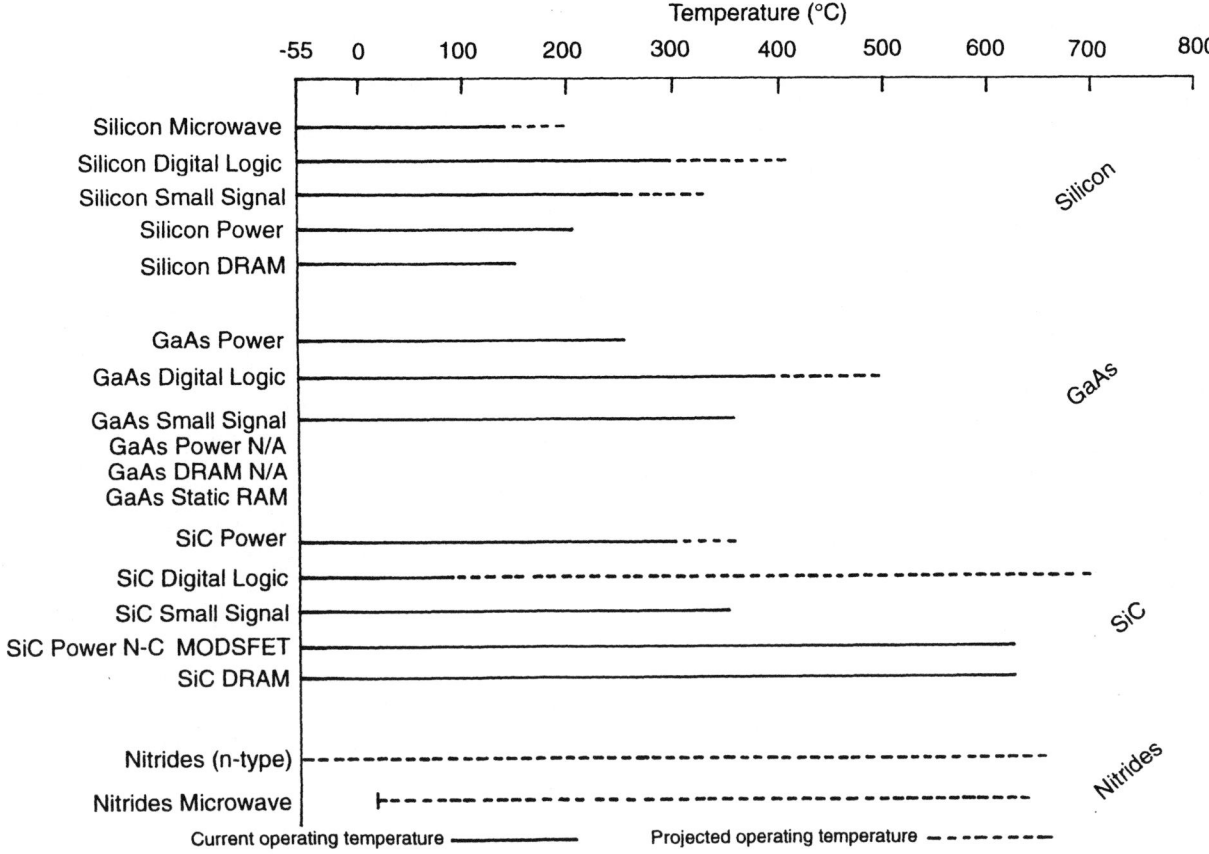

FIGURE 3-7 Operating temperatures for different devices per material.

by the changes in leakage currents, gain, and threshold voltages caused by high temperature. Microwave devices have been operated at 300 °C (Wurfl et al., 1994), and the extrapolation to 350 °C operation reflects again the more constraining effects of temperature on these devices, compared to small-signal analog devices.

DRAMs, in the conventional sense, are not made using GaAs. As for power devices, conventional metal-insulator semiconductor field effect transistors (MISFETs), popular power devices in silicon, are not possible in GaAs, due to the high interface recombination velocity. Other power devices, including the thyristor-based devices, have not yet been attempted, even with the excellent development of other device types and integrated circuits in this materials system.

Although the wide bandgap materials still represent nascent technologies, there have already been demonstrations of devices that have exhibited DC transistor operation at considerably higher temperatures than available through silicon or GaAs devices. SiC transistors have shown operation at temperatures as high as 650 °C (Palmour et al., 1991), and they are projected to operate at still higher temperatures. Based on material fundamentals, operation to over 1000 °C should be possible. There is as yet little information on SiC digital logic. Xie et al. (1994) demonstrated some SiC enhancement-mode n-channel metal-oxide semiconductor (NMOS) circuits to 300 °C, which included simple gates, latches, and flip-flops. Based on the demonstrated transistor characteristics, and on material fundamentals, operation to at least 650 °C should be possible. D.M. Brown et al. (1994) have demonstrated operational amplifier operation to 350 °C. The extension to 550 °C is based on estimates similar to those for silicon and GaAs. Similarly, the same estimates apply to the expected performance of power devices at elevated temperatures. The high-frequency results for SiC are currently not available. Thus the committee was unable to make an accurate appraisal of its microwave potential. The lack of a SiC CMOS for analog could restrict potential device applications, however.

Device work in the nitrides is in the very early stages; however, the results are already very exciting. The current devices are made in epitaxial layers with very high defect densities. Nevertheless, reasonable transistor action has already been achieved to 350 °C (Binari et al., 1994). Based on fundamentals, operation to over 1000 °C should be possible. Although it is too early to give an accurate estimate of high-temperature operation of most applications (e.g., digital logic small-signal analog, power, and DRAM), if the estimates based on operation of transistor action are reasonably accurate, operation to very high temperatures should be possible. The projected operating temperatures in Figure 3.7 are based on estimates similar to those for silicon, GaAs, and SiC. As for transistor operation, the results for microwave devices at this early stage of materials development are very encouraging. Maximum frequencies of oscillation as high as 35 GHz have already been reported (Khan et al., forthcoming).

Figure 3.7 should serve only as an approximate basis of comparison, as is true for the figures of merit, representing informed extrapolations. To uncover the true limitations or capabilities will require further experimentation. The intention is to more clearly focus on desired operating temperatures and the concomitant most promising device technologies in those temperature regimes.

4

Generic Technical Issues Associated with Materials for High-Temperature Semiconductors

For the high-temperature semiconductors being considered in this report, process technologies must be assessed in addition to performance effectiveness. The existence of processing technology that is compatible with high-quality materials will determine whether a device and circuit technology can be built from these materials.

This chapter discusses the most important materials-dependent processing technologies required for the production of high-temperature semiconductors. The following issues are considered: (1) electrical contacts, (2) doping and implantation, (3) gate oxides and insulators, (4) etching, (5) defect engineering and control, (6) yield, and (7) device reliability. The committee expects more focus on these areas as the high-temperature device technologies mature and as more long-term reliability studies at elevated temperatures take place.

ELECTRICAL CONTACTS

The requirements for high-quality contacts for high-temperature electronic devices are similar to their low-temperature counterparts, but these contacts may be more prone to degradation at their elevated temperatures of operation. Therefore, studies on the mechanisms of intermetallic and metal semiconductor reactions, interdiffusion reactions, and electrical changes as a function of time will be critical to projecting device stability and reliability. Finding a means of forming effective, reliable contacts to high-temperature semiconductors will obviously have important consequences to other device applications of these wide bandgap materials.

This section focuses on ohmic contacts to SiC and GaN, and Schottky contacts to SiC. Of the wide bandgap semiconductors, research in electrical contacts to these materials is currently the most mature. Although progress with GaN has been encouraging, ohmic contacts with low contact resistance remains a significant issue that needs to be addressed for the nitrides and diamond. Some of the high-temperature metallization issues for ohmic contacts are addressed since the initial choices of metals must begin by addressing the high-temperature issues. Extensive testing and stressing at high temperatures is only in initial stages, however.

Schottky Contacts to SiC

Most transition metals form good Schottky contacts on n-type (α- and β-) and p-type (α-) SiC with Schottky barrier heights (SBHs) typically greater than or equal to 1 eV. Positive correlations that were less than 1 eV between the SBHs and work functions of several metals were found in a number of studies (Waldrop et al., 1992; Porter et al., 1993, 1995a, b; Waldrop and Grant, 1993). These results indicate a partial pinning of the Fermi level in SiC. Low leakage currents at room temperature for contacts on both n- and p-type 6H-SiC have been common, while the characteristics of contacts on β-SiC films have been very dependent on the quality (i.e., defect density) of the β-SiC films. Breakdown voltages greater than 1,100 V for gold diodes on the carbon-terminated surface (Urushidani et al., 1993) of n-type 6H-SiC have been achieved. Recently, Schottky diodes in pre-production packaging have been demonstrated with breakdown voltages greater than 1,200 V.

Ohmic Contacts to SiC

The combination of the wide bandgap and surface states in SiC make the fabrication of ohmic contacts to this material, especially p-type SiC, exceptionally difficult

TABLE 4-1 Selected Ohmic Contacts to n-Type 6H-SiC and Measured Contact Resistivities at Room Temperature

Contact Metallization	Carrier Concentration (cm^{-3}) in 6H-SiC	Annealing Conditions	Contact Resistivity ($\Omega \cdot cm^{-2}$)	Reference
TiW	4.7×10^{18}	600 °C/5 min	7.8×10^{-4}	Crofton et al., 1991
Titanium	1×10^{20}	none	$<2 \times 10^{-5}$	Alok et al., 1993
Nickel	$7–9 \times 10^{18}$	950 °C/2 min	$<9 \times 10^{-6}$	Crofton et al., 1994
Ni/3C-SiC	$1–2 \times 10^{18}$	1000 °C/30 s	$1.7–6 \times 10^{-5}$	Dimitriev et al., 1994

SOURCE: Porter et al. (1995b).

to achieve. Most ohmic contacts rely on high doping concentrations in the SiC and must be annealed at temperatures between 800 °C and 1300 °C (Porter et al., 1995a, b). Tables 4-1 and 4-2 list selected ohmic contact metallizations and their corresponding contact resistivities on n- and p-type 6H-SiC, respectively. Additional ohmic contacts are shown in Table 4-3. Annealed nickel and aluminum have been the most common ohmic contact metallizations for n- and p-type (α- and β-) SiC, respectively. Because of the very high thermodynamic driving force for oxidation of aluminum, annealing aluminum, or aluminum alloys can result in the formation of an insulating oxide layer.

The fabrication of ohmic contacts that have low contact resistivities ($<10^{-5}$ $\Omega \cdot cm^2$) and are thermally stable on SiC will be one of the most critical issues for the advancement of SiC devices. In addition to reducing the Schottky barrier, the reactions at the metal/SiC interface and their kinetics must be considered if SiC devices are to be operated at high temperatures.

Ohmic Contacts to GaN

Table 4-4 summarizes some of the state-of-the-art contacts made to GaN. Generally, good ohmic contacts have been made to both n- and p-type GaN. SBHs are found to be dependent on the metallic work function, indicating that the surface Fermi energy of GaN is unpinned, in contrast to both SiC and ZnSe (and most other compound semiconductors). Morkoc et al. (1994) speculates on the nature of the low-resistance contacts to GaN. Two possible mechanisms for low-resistance contacts are discussed: low-barrier Schottky contacts coupled with intermediate or graded bandgap interfaces and tunneling.

DOPING AND IMPLANTATION

Finding suitable shallow dopants for large bandgap semiconductors is one of the major limitations to the development of these materials for device applications. Doping has been achieved by co-deposition during growth of the film itself, or by ion implantation. Both of these technologies have been employed for all the large bandgap semiconductor materials with greater or lesser success, depending on the particular system. Diffusion doping has been attempted in some cases, but this approach has been found to be unacceptable for device fabrication.

Doping to achieve both n- and p-type materials in SiC has been shown to be easier, for example, than in either GaN or ZnSe. For SiC, nitrogen has been predominantly used as the n-type dopant, while aluminum is generally used as the p-type dopant. With respect to doping by implantation, SiC, particularly 6H-SiC, has been the most widely studied of all large bandgap semiconductor materials. Methods for both n- and p-type doping of Group III nitrides are required. As improved quality materials become available in GaN and AlN, similar studies are anticipated.

Doping of SiC

Nitrogen is the most frequently used n-type impurity, while aluminum is most common for p-type doping in

TABLE 4-2 Selected Ohmic Contacts to p-Type 6H-SiC and Measured Contact Resistivities at Room Temperature

Contact Metallization	Carrier Concentration (cm^{-3}) in 6H-SiC	Annealing Conditions	Contact Resistivity ($\Omega \cdot cm^2$)	Reference
Aluminum	1.8×10^{18}	700 °C/10 min.	1.7×10^{-3}	Crofton et al., 1991
Al-Ti	2×10^{19}	1000 °C/5 min.	1.5×10^{-5}	Crofton et al., 1993
Al/3C-SiC	not reported	950 °C/2 min.	$2-3 \times 10^{-5}$	Dimitriev et al., 1994

SOURCE: Porter et al. (1995b).

CVD (chemical vapor deposition). Dopants may be introduced either during epitaxy or later using ion implantation. For CVD, nitrogen and triethylaluminum (TEA) have proven to be suitable dopant source gases for n- and p-type doping, respectively. When n-doping is introduced during growth, carrier concentrations as high as 1×10^{19} cm^{-3} can be realized. Ion implantation with subsequent argon annealing has yielded electron concentrations as high as 3×10^{19} cm^{-3} at an n-dose of 5×10^{20} cm^{-3} (D.M. Brown et al., 1994; Clarke et al., 1993).

Acceptor, p-type doping is a recognized problem in SiC, although considerable progress has been made. All of the acceptor impurities thus far investigated, namely aluminum, boron, gallium, and scandium, form deep levels, are difficult to activate, and generally require a high-temperature anneal. The depth of the acceptor levels also leads to the hole concentration varying quite strongly with respect to temperature, considerably complicating device design and operation. Aluminum is somewhat difficult to incorporate into the SiC lattice and high carrier concentrations are difficult to achieve. Researchers at Kyoto University have obtained p-type carrier concentrations in the 10^{19} to 10^{20} cm^{-3} range, using TEA in a CVD process on the silicon face of 6H-SiC. In contrast, growth on the carbon face permitted only 2×10^{18} cm^{-3} p-type doping. The carrier concentration could be easily controlled down to the low $p = 10^{16}$ cm^{-3} range. On the upper end, the observed hole concentration became nonlinear as a function of TEA flow above 10^{19} cm^{-3}.

In general, background nitrogen causes unintentionally doped crystals to be n-type. In the best 6H-SiC samples (<10^{14} cm^{-3} nitrogen), background carrier concentrations in the mid-10^{14} cm^{-3} range have been achieved. Further improvements should be possible as sources of nitrogen contamination are eliminated.

A review of the optical and electrical properties of doped SiC has been published by Pensl and Choyke (1993). One important point made by the authors is that dopants can occupy either hexagonal or cubic sites in the more complex SiC polytypes. These different environments give rise to different binding energies and care must be taken when deconvolving the separate contributions from Hall data. Pensl and Choyke showed that the relative abundance of the various nitrogen dopant levels corresponded to the ratio of available binding sites. For 4H-SiC, the hexagonal (h) and cubic (k) binding energies were measured to be 45 meV and 100 meV, respectively. A level ratio of 2:1 was found in 6H-SiC, reflecting the fact that two-thirds of the sites have cubic bonding. In 6H-SiC Hall measurements, the measured ionization energy of the hexagonal site was 85.5 meV, while the cubic sites were 125 meV. The experimental resolution was insufficient to resolve the two separate cubic donor energy levels. In 3C-SiC, a value of 48 meV was determined. Typical compensation values were one to two orders of magnitude below the observed electron concentration. Similar measurements for aluminum-doped SiC yielded an acceptor ionization energy of roughly 200 meV for each of the three most common SiC polytypes. These values are all smaller than those measured optically due to a reduction in the average electron energy when donor spacing is small.

In commercial SiC technology, ion implantation plays a major role. Due to the excellent stability of SiC, the material lends itself well to high-temperature annealing for implantation-related damage removal. Marsh and Dunlap

TABLE 4-3 Additional Ohmic Contact for SiC

Metal	Post-Treatment	Contact Resistivity ($\Omega \cdot cm^2$)	Description	Reference
Nickel	High temperature		Ohmic to source/drain limited to 300 °C or less	McGarrity et al., 1992
Aluminum	High-temperature anneal		Gate	McGarrity et al., 1992
Gold, platinum, titanium, hafnium, cobalt to n/p-type SiC	> 450 °C	10^{-3} to 10^{-4}		Morkoc et al., 1994
To heavily doped p-type		10^{-6}		
WSi_2		4×10^{-4}	Delaminates at 600 °C	Morkoc et al., 1994
Au/Ta/SiC	900 °C in air	1×10^{-5}		Morkoc et al., 1994

(1970) characterized the first ion-implanted SiC junctions in 1970 that were formed by implanting n-type dopant into a p-type substrate at room temperature. More-modern approaches utilize a process in which the target material is heated during the implantation (Ghezzo et al., 1992). Ghezzo et al. (1993), at GE, have demonstrated improved diode characteristics using boron-implanted 6H-SiC at high temperature with a post-annealing. A summary of the needs for ion implantation for SiC devices has been presented by D.M. Brown et al. (1994).

Doping of GaN

A review of the doping of GaN is given by Morkoc et al. (1994). Unintended doping of GaN, as well as AlN, results from an n-type background concentration. With improved crystal growth techniques, background concentrations for GaN have recently been reduced to as low as 10^{16} cm^{-3}. For example, Nakamura et al. (1992) have reported GaN bulk mobility, μ = 600 to 1,500 cm^2/V·s at 300 and 77 K, respectively, in an undoped sample with n = 4×10^{16} cm^{-3}. Nitrogen vacancies are considered to be the most likely candidate as a donor site. Control of the nitrogen overpressure appears to be critical to influencing this type of defect. A resurgence in activity has occurred with the recent observations by Amano et al. (1989) of low-energy electron-beam activation of magnesium-doped GaN to produce p-type GaN. Nakamura et al. (1992) have improved on these results by using low-energy electron-beam irradiation during the implantation. They also discovered that annealing the magnesium-doped GaN at 700 °C in nitrogen produced equally good p-type GaN. The process was reversible with NH$_3$ anneal where hydrogen is found to be a compensating agent. Molecular-beam epitaxy (MBE) processing that is free of hydrogen has been found to induce p-type conducting in the as-grown state. Numerous other dopants have been used in an attempt to produce p-type GaN, with zinc as the most effective p-type impurity. A list of these are presented in a review article by Morkoc et al. (1994).

In addition, isolation regions have been produced by proton ion implantation. Energies and doses were unspecified except to say that the energies and doses were selected to produce uniform compensation across a 1-μm-thick film of 1×10^{17} cm^{-3} n-type GaN epitaxial layer.

Doping of AlN

There has been little work in the area of doping of AlN. A persistent contaminant of AlN is oxygen. It is clear that major improvements and much more work is

TABLE 4-4 Ohmic Contacts for GaN

Metal	Post-Treatment	Contact Resistivity ($\Omega \cdot cm^2$)	Description	Reference
Titanium 25 Å/ Gold 1,500 Å	250 °C for 30 s	7.8×10^{-4}	Ohmic to source & drain MESFET	Khan et al., 1993a, b
Silver		Schottky	Gate metal No other info.	Khan et al., 1993a, b
Ti/Al/n-GaN	900°C for 30 s	8×10^{-6}	Good ohmic	Lin et al., 1994
Gold and Au/Ni/p&n-GaN		Contact resistance not reported, but reasonable contact resistance deduced from operating voltage of 4 V at 20 mA		Nakamura et al., 1991
Al/n-GaN		10^{-4}		Foresi and Moustakas, 1993
Au/GaN		10^{-3}		

required to produce useful device-quality materials. Ion-implantation doping of AlN is virtually unknown, and work in this area will require quality substrate materials that are only recently becoming available.

Doping of Diamond

Boron is the only universally recognized acceptor impurity that can be controllably introduced into diamond. Until recently, only a small portion of the boron in diamond was electrically activated. It can now be introduced and nearly 100 percent electrically activated by a series of implantation processes to concentrations of 1×10^{19} cm^{-3}. The same investigator stated that he has activated phosphorous in diamond at 80 MeV by a similar procedure (Prinz, 1994).

GATE OXIDES AND INSULATORS

This section examines issues related to the growth of suitable gate oxides and insulating layers for field effect devices. These are either thermally grown oxides, where appropriate, or epitaxially grown higher bandgap layers, where stable, high-quality oxides are difficult to achieve. Not included in this discussion are films that are deposited to provide electrical isolation (e.g., silicon oxides or nitrides). Although there may be fundamental problems with the identification of materials that will withstand the high-temperature environment for which the associated devices are intended, the committee regards these issues as secondary in the sense that they are not intrinsically related to a particular material choice.

Gate Oxides and Insulators for SiC

SiC benefits from its amenability to thermal oxidation to form the well-understood, characterized, and utilized SiO_2. The general consensus is that the oxidation of SiC follows that of the well-established silicon oxidation process. The total oxide thickness, c, can be estimated by:

$$c^2 + Ac = B(t+t_0),$$

where t is the duration of the oxidation process, A and B are temperature-dependent rate constants, and t_0 is a constant that depends on the initial state of the surface (Deal and Grove, 1965). As for silicon, oxidation rates and oxide quality will vary according to the SiC polytype, the crystal orientation, defect density, doping level, and the nature of the oxidation conditions (i.e., whether carried out under wet or dry ambient conditions). For example, Laukhe et al. (1981) found that the 3C, 4H, 6H, and 15R polytypes oxidized at different rates on the silicon face, with 3C oxidizing approximately 20 percent more rapidly than 6H. The oxidation of the 6H and 3C polytypes on the silicon and carbon faces has been characterized, contrasting the differences brought about by either wet or dry oxidation and determining values for the rate constants A and B in the above equation (Powell et al., 1991; Petit et al., 1992). Significant differences in the oxide growth rates of polytypes on the carbon face, as well as the silicon face, were found. The differential oxidation rates were used to provide a map of the polytype distribution for the silicon and carbon faces of SiC films. Little difference in oxidation rates of the polytypes were found when the silicon face was oxidized in a wet ambient or when the carbon face was oxidized in a dry ambient; conversely, the 3C polytype oxidizes more rapidly in dry oxidation of the silicon face and the 6H more rapidly in wet oxidation of the carbon face. These differences are obviously of critical importance in the fabrication of MOS-quality devices. However extensively characterized, the quality of the SiC/oxide interface will still require considerable improvement before a reliable device technology can be assured. For example, in Singh and Rys' (1993) comparison of wet and dry oxidation of n-type, silicon-face 6H-SiC, capacitance–voltage measurements revealed the necessity for a post-oxidation in argon. Otherwise, the capacitance–voltage curves showed no accumulation after the first trace was taken. Under dark conditions, inversion did not occur—perhaps because of the lack of minority carriers—due to the large bandgap of the SiC. Other work has found that sweeping from accumulation to inversion in real time required heating the substrate to 860 °C (Morkoc et al., 1994). The extensive characterization of oxide samples revealed interface trap states between mid-10^{11} cm$^{-1}\cdot$eV^{-1} and 10^{12} cm$^{-1}\cdot$eV^{-1}, with lower trap densities for dry compared with wet oxidation, but comparable emission time constants ($\sim \mu$s). The nature of the traps may differ depending on whether the oxidation is wet or dry, however, these values of interface states density has found corroboration in other work on n- and p-type samples (Morkoc et al., 1994). Finally, a good dielectric strength of 6×10^6 V/cm has been measured for thermally grown oxide on 3C-SiC (Fung and Kopanski, 1984). Future approaches to form a device-quality oxide may involve the use of alternative oxidation sources, such as N_2O.

Gate Oxides and Insulators for the Nitrides

Based on the wealth of experience and research on compound semiconductors such as GaAs and InP, it is not expected that a suitable oxide-based device technology will be developed in these materials. However, the GaN/AlN system may have particular advantages in formation of metal insulator semiconductor (MIS) and electrically insulating structures capitalizing on the reasonable lattice-match of the AlN and GaN (2.4 percent), and the differences in bandgap (6.2 eV for AlN, 3.4 eV for GaN). The higher bandgap AlN or $Al_xGa_{1-x}N$ can then serve as an insulator, playing the same role as $Al_xGa_{1-x}As$ in GaAs/AlGaAs semiconductor-insulator-semiconductor field effect transistors (SISFETs), and the device structure would closely approximate that shown in Figure 4-1. To give further impetus to this approach, previous measurements have indicated that the resistivity of unintentionally doped $Al_xGa_{1-x}N$ increases rapidly with increasing aluminum mole fraction, becoming insulating at about x = .20 (Figure 4-2), minimizing the possibility

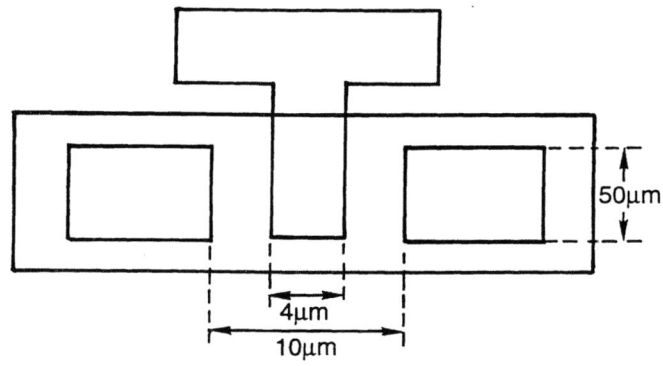

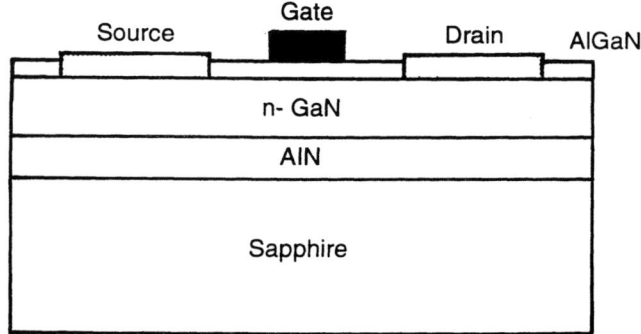

FIGURE 4-1 Schematic of the device structure for a AlN/Al_xGa_{1-x}N SISFET. SOURCE: Morkoc et al. (1994).

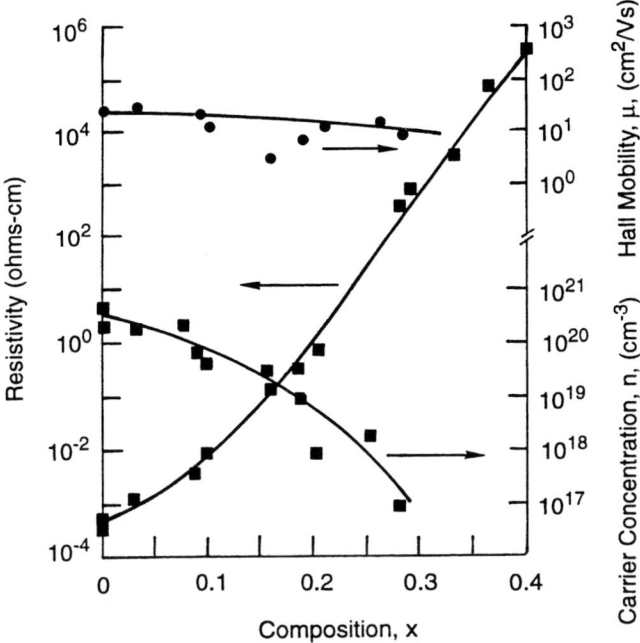

FIGURE 4-2 Increase in resistivity of unintentionally doped $Al_xGa_{1-x}N$ with increasing aluminum mole fraction. SOURCE: Morkoc et al. (1994).

of leakage through the AlN insulator layer. Critical to the feasibility of the approach will be the nature of the interface between GaN and AlN (or GaAlN), as well as the quality of the insulator material itself. Optimism for a high-quality interface is built upon the already demonstrated high-quality operation of an AlGaN/GaN modulation-doped field effect transistor (Khan et al., 1993a). Modulation of the GaN/AlN interface can be inferred from the ability to modulate the two-dimensional electron gas and achieve complete pinch-off of this transistor.

Gate Oxides and Insulators for Diamond

No material-compatible oxide or insulating layer has been identified for the diamond-based technology thus far. Device realizations have generally comprised MESFET-like structures. The production of MOSFETs using CVD oxides has also been attempted.

ETCHING

Pattern transfer by etching is a critical component of any device and circuit fabrication scheme. For certain stages in the fabrication sequence, such as etch cleaning of the substrate surface, or etch delineation of patterns having large dimensions (tens of microns), wet chemical etching is the preferred process. Many of the materials under discussion here, such as SiC or diamond, are relatively inert to general chemical etchants. However, constraints of high-resolution features, control over etched profiles, and etch-rate and process reproducibility have generally led to dry etching techniques as the choice for manufacturing-compatible processes. In this regard, dry-etch chemistries and processes have already been applied and are meeting with reasonable success. The manufacturing-related issues of uniformity and control will have to be further explored, as is being done even for the silicon-based technologies. Furthermore, although the initial applications of dry etching to the wide bandgap materials have met with success, much more work remains in order to delineate the basic mechanisms, the rate-limiting steps, the quality of the etch surface, the possibility of etch-induced damage, etc.

Etching of SiC

Successful reactive-ion etching (RIE) has been carried out on SiC, primarily incorporating the same chemistries that have been established and proven successful for silicon-based technology (i.e., the use of fluorine-containing gases that reacts with silicon to form the volatile SF_4). Pan and Steckl (1990) have utilized SF_6, CF_3Br, and CHF_3 mixtures with oxygen and generally achieved modest etch rates of a few hundred angstroms per minute. A factor that differs from conditions in which most silicon RIE is carried out is the generally large component of oxygen utilized (35-90 percent by flow), thought to be important in removal of the carbon through generation of the volatile product, CO_2. Chlorine (Balooch and Olander, 1992) and chlorine-containing gases, such as CCl_4 (Lo and Huang, 1992), have also been utilized, as well as CF_4/O_2 (Padiyath et al., 1991) and NF_3/O_2 (Brown, 1993). Although the etch rates obtained range from 50-200 nm/min and are comparable to those found for RIE of silicon under a very fluorine-rich etch condition (SF_6 + 35% O_2; Pan and Steckl, 1990), silicon was found to etch about an order of magnitude more rapidly than SiC.

Generally, the strong bonding of SiC makes it a difficult material to etch by wet chemical means at reasonably low temperatures. However, Shor et al. (1994) have found that etching using laser illumination of a sample immersed in an HF-based solution, has produced etch rates well in excess of 1 μm/min. Such a photoelectrochemical etching technique may have applicability to certain fabrication requirements, such as deep etches that may be needed for sensor fabrication.

Etching of the Nitrides: GaN and AlN

Dry etching of the nitrides has benefited from the previous extensive research on gallium and aluminum-containing semiconductors, in a manner similar to SiC and silicon. With respect to the periodic table of elements, the successes demonstrated thus far indicate that it is the Column III elements that provide the major constraint to the RIE of the nitrides. Generally, chlorine-based chemistries have been applied to the etching of GaN and AlN, comprising gases such as CCl_2F_2, BCl_3 (Pearton et al., 1993; Lin, 1994), $SiCl_4$ (Adesida et al., 1993), and $SiCl_2F_2$ (Morkoc et al., 1994). Adesida and colleagues achieved a GaN etch rate of 600 Å/min using $SiCl_4$ but only at higher voltages, suggesting a rather low chemical enhancement in this process. Similarly, Pearton et al. utilized electron cyclotron resonance radio frequency etching of the nitrides in CCl_2F_2 and BCl_3, at relative low bias (<200 V), low pressure, and low microwave power. The etch rates were approximately 200 Å/min, which is modest even at these low pressure and bias conditions. However, Lin et al. (1994) used the BCl_3 chemistry and were able to obtain greater than 1,000 Å/min etch rate. Temperature-dependent etch-rate studies would help to more clearly reveal the rate-limiting steps in current dry-etching approaches and would also give a better idea of the ultimate etch rates and profiles achievable. In a manner similar to forming selective GaAs to AlGaAs etch processes, the addition of excess fluorine to a gas composition will provide high selectivity of etching GaN with respect to AlN (Pearton et al., 1993). The etch anisotropies obtained for all methods used have been excellent.

Etching of Diamond

Diamond is commonly regarded as an inert material that cannot be etched by boiling acids or bases. Nevertheless, diamond etch rates as high as 200 nm/min were reported by Efremow et al. (1985), using ion-beam-assisted etching that utilized 2 keV Xe^+ ions with a flux density of 1 mA/cm² in the presence of NO_2 reactive gas. Although the average pressure of the chamber was 2×10^{-4} Torr, the equivalent reactive flux of NO_2 on the diamond corresponded to 10^{-5} Torr, presumably responsible for the impressive etch rates that were achieved, with etch-rate selectivities to aluminum masks at a ratio of 20:1. Other methods that have been used to etch diamond are (1) the use of kinetic energy beams of oxygen or oxygen-containing molecules or radicals and (2) electrolytic etching. Electrolytic etching is generally limited to removal of defect-ridden or otherwise conducting regions of diamond, however.

DEFECT ENGINEERING AND CONTROL

Every materials system presents new challenges in defect engineering to produce and maintain desired performance. High-temperature materials are robust at 300 °C, but may be vulnerable at the high process-temperatures required. This section evaluates crystal perfection, processing, and electrical performance for diamond, SiC, and GaN. The purpose is to identify fundamental limitations and focus areas for precompetitive research and development leading to commercialization.

Semiconductor materials composed of low atomic number elements possess strong covalent bonds. As a result they exhibit large energy gaps, high elastic moduli, high phonon frequencies, high thermal conductivity, and high melting points. These properties, which can be exploited for high-temperature electronics, have consequences in processing and perfection. The formation energy of simple point defects (vacancies and interstitials), $E_f(V)$, $E_f(I)$, can be estimated to first order as the heat of vaporization, ΔH. This energy represents creation of a bulk defect from a surface source, and it is typically valued at twice the bond energy.

Thus, a lower bound for E_f is twice the energy gap, Eg. With Eg = 3 eV, E_f is expected to be very large. These large values preclude doping by diffusion and present a bottleneck for defect control (e.g., during oxidation and contact formation). If process temperatures do not exceed one-half the melting temperature, stoichiometry (the equivalency of conjugate defects on each sublattice of a compound) can be maintained, dislocation motion is limited, and the incorporation of unwanted contamination is reduced. However, Larkin et al. (1993) have shown that dopant incorporation is dependent on the silicon/carbon ratio in nonequilibrium growth.

If electrical properties scale with the energy gap, most nondopant impurities should have deep energy positions, with energy of the trap level $E_T > 1$ eV, and therefore they should not contribute to leakage current generation in junctions at 300 °C. Large concentrations of these trap states, however, could result in semi-insulating behavior. The compensation level (i.e., the difference between the number of acceptors and the number of donors, $N_A - N_D$) must be defined in device-quality material, and a theory and catalog of defect and impurity energy positions must be established.

Since directional covalency dominates bonding, local relaxation from fourfold coordination can greatly reduce defect and impurity incorporation energies. For instance, the nitrogen donor in diamond lowers its symmetry from a substitutional site to produce a deeper state than predicted by simple theory. Davies (1994) has recently reviewed the properties of defects and impurities in diamond. Nitrogen in 6H-SiC shows persistent photoconductivity indicative of this relaxation. The difficulty in p-type doping of these wide bandgap materials may similarly be related to the preference of Group III atoms for threefold coordination. In general, low atomic number elements from groups III, IV, and V act as electronically passive terminations for surfaces and defects. Phosphorus-containing compounds exhibit less sensitivity of device performance to defects and more passive surfaces.

The intrinsic carrier mobility and minority carrier lifetime in these materials must be characterized and understood. In the mid-1950s, these properties were well-defined for silicon, providing a benchmark for material quality. This foundation does not yet exist for high-temperature semiconductor materials. SiC, the most developed material of the group, has a reported minority carrier lifetime of 10^{-9} s, and carrier mobilities that increase with activated behavior above room temperature. But the issue arises whether these properties are intrinsic or reflections of high defect densities that are not yet characterized. Swedish researchers have reported minority carrier lifetimes in 6H-SiC of 0.45 μs (Kordina et al., 1995).

Passivation of surface conduction and recombination is essential to device function. It is essential to measure interface properties at the temperature of operation. Kang et al. (1993) have shown that quasi-static C-V measurements at 20 °C can underestimate by an order of magnitude the density of interface states of 6H-SiC/SiO$_2$ that are active at 240 °C. Performance at 400 °C cannot be predicted. The porosity, state of stress, and oxidation mechanism of SiC must be understood. Other issues also remain: (1) there is a factor of 10 higher oxidation rate on (111) carbon-terminated versus silicon-terminated surfaces that has not yet been explained, (2) the true density of interface states is unknown, and (3) the failure mechanism for gate-oxide breakdown at operational temperatures is not known.

The quality of critical insulators (e.g., gate oxides in MOSFET technologies), and hetero-interfaces (e.g., oxide-to-SiC or GaN-to-AlN) will be important determinants of not only device performance but also device reliability. Defects in the oxide layer and at the oxide-semiconductor interface may arise through propagation of defects from the semiconductor substrate, may be correlated with the manner of oxidation (e.g., whether carried out in a wet or dry ambient and at a particular temperature), or may be introduced by subsequent processing (e.g., gate recess etching performed by reactive-ion etching). As current silicon technology has shown, these problems are aggravated as the oxide layers have been made thinner in order to appropriately scale device dimensions down to achieve denser, faster circuits with higher functionality. The reliability of heterojunction devices such as SISFETs will similarly depend on the nature of the interface; interface roughness and interface state densities will affect device mobilities and increase device noise, as well as long-term performance.

YIELD

High yield in device and circuit manufacturing requires high-quality substrates and wide process margins. Typical SiC contains 20-100 micropipe voids and 10^4 to 10^5 cm^{-2} dislocations. GaN is grown epitaxially with approximately 10^{10} cm^{-2} dislocations on high misfit substrates such as sapphire. The electrical properties of these dislocations have yet to be understood. They are of no apparent consequence in GaN blue light-emitting diodes, but these dislocations may limit yield for other materials or applications by providing fast diffusion paths for unwanted impurities during processing or by acting as junction shorts. Micropipes act as premature breakdown sites in SiC junctions (Fazi et al., 1993). Material screening should evaluate the role of microplasma sites in premature breakdown at operational temperatures.

Dopant uniformity is essential to process control. Implant activation for aluminum and boron in SiC requires further development (aluminum exhibits < 1 percent activation). Activation is dependent on the perfection of the host material. Nonuniformity across a wafer suggests that spatial mapping of defect densities (X-ray topography, photoluminescence, etc.,) could provide a valuable characterization matrix in materials development. Of course, control of dopant compensation in the 10^{14} cm^{-3} range is a necessity.

The role of transport anisotropy must be evaluated in the development of a planar integrated circuit technology. The mobilities and effective masses are very anisotropic for 6H- and 15R-SiC. By contrast, 4H-SiC is more isotropic.

Etch and metallization processes can create defects and yield limits. The absence of a complete wet-etch technology for high-temperature semiconductor materials means that RIE is the default process. Defect introduction and stability for each RIE application must be evaluated. Contact formation and interfacial reactors can modify band-edge offsets and, hence, contact resistance and heterojunction device performance. These reactions and process response surfaces must be understood. Alternative approaches, such as layer bonding and buffer-layer engineering, should be pursued. The small lattice constants of the high-temperature semiconductors have no available match among the mature substrate materials technologies.

In summary, process yield is dependent on the critical defect density and the chip area. The next phase of research and development on high-temperature semiconductors must identify the critical defects and improve materials and processes to reduce their density to economically tolerable levels.

DEVICE RELIABILITY

The expected applications for high-temperature electronics (e.g., aircraft, automotive, and power systems) require high reliability. However, the temperatures of operation are typical of accelerated aging conditions for current systems. The purpose of this section is to identify potential failure mechanisms and suggest areas of research for remediation. The current design approach is to extrapolate low-temperature concepts to higher temperatures. However, new failure modes may exist in this unexplored region with new materials systems. Detection of these new modes is a critical subject of research.

High-temperature systems will cycle between room temperature and higher operation temperatures. The mechanical design of these components must include differential thermal expansion and consequent fatigue tolerance as factors in materials selection.

Current solder alloys will melt by 300 °C. New alloys with both mechanical strength and systems compatibility must be found. Polymer dielectrics for encapsulation and printed circuit boards must similarly be redesigned. Integral ceramic packages are a favored replacement.

Electromigration of interconnect lines is thermally activated, as well as current density dependent. For silicon-on-insulator (SOI) applications, aluminum (T_M = 600 °C) must be replaced. Tungsten and copper are likely candidates.

Thermal management is particularly critical for high-power applications at high temperatures. Silicon bipolar devices exhibit "thermal runaway" at 200 °C due to free carrier generation. SOI limits rise to 350 °C due to the smaller volumes of silicon.

Gate oxide and hot carrier degradation are cumulative effects. Their temperature dependence is controversial. The role of temperature in aging and operation must be understood.

Contact degradation is a serious problem for III-V materials at high temperature. Pt-GaAs impact avalanche transit-time (IMPATT) diodes degrade as the platinum metallization reacts to form $PtAs_2$ and PtGa. Also, defect migration from the contact can compensate the doping level and change the device avalanche voltage. Diffusion barrier layers and stable heterogeneous systems must be designed.

In laser devices, high injected current densities lead to degradation by "dark-line-defect" formation (electronically stimulated dislocation motion). The high elastic moduli of diamond, GaN, and SiC inhibit dislocation motion. In addition, the low atomic number elements tend to create electronically passive defects and interfaces. Dislocations appear to be electrically benign in GaN. These materials should display better operational reliability than GaAs, InP, and ZnSe.

In summary, the strategy for reliability assurance is (1) re-evaluation of known failure mechanisms in a high-temperature applications context and (2) identification and remediation of new failure modes in these new materials systems and environments.

5

High-Temperature Electronic Packaging

Preceding chapters of this report have presented the status and potential of the use of various semiconductor devices for high-temperature applications. Depending on the given application, the operating temperature of these devices varies from 150 °C (automotive electronics) to 1000 °C (a high-temperature interconnect for a thermionic integrated circuit; Fendrock and Hong, 1990). Adequate packaging has to be provided to build functional modules based on these devices, however. The primary issues to be considered in high-temperature electronic packaging are: (1) characterizing materials and their interactions at high temperatures, (2) minimizing mechanical stresses caused by thermal expansion mismatches, (3) providing a suitable path for heat dissipation, and (4) providing environmental protection. The first two items are of particular importance as the upper temperature limits of operation are increased. The overall goal of packaging is to ensure reliable operation of semiconductor devices.

The selection of the packaging method used is based on cost/performance tradeoff decisions, complexity of the devices involved, system requirements, and the operating temperature range. There are many different packaging and interconnection schemes that can be successfully applied to overcome the shortcomings of conventional packages for high-temperature operation.

A successful package design will satisfy all given applications and system requirements. For example, microwave packaging will have to meet additional requirements such as impedance matching, low dielectric loss at microwave frequencies, low sensitivity of dielectric and conductors to temperature changes, and low capacitance of interconnect to the backside of radio frequency ground plane. One approach to providing hermetic electronic packaging suitable for operation at higher temperatures is to rely on existing packaging technology such as either the use of single metal packages with electrical feedthroughs isolated by glass beads or the use of multilayer ceramic packages (single or multichip). Both types of packages have been successfully used for high-temperature electronics. Another approach is to seek improved packaging materials and designs that better match the needs of high-temperature electronic systems.

This chapter focuses on the off-the-shelf packaging technologies used in both the first level of high-temperature electronic-system packaging, usually defined as the chip packaging, and the second-level packaging, which is defined as package-to-board interconnections. Since some large hybrid microcircuits and multichip modules closely resemble both first- and second-level packaging and because the high-temperature electronic boards are often ceramic with thick- or thin-film metallization, they share the same materials and assembly processes and are considered here to be the same high-temperature electronic-packaging technology (Palmer and Heckman, 1978; Palmer, in press).

CHIP PACKAGING

High-temperature electronic-packaging technologies have been developed at least five times over the last four decades (Palmer, in press): vacuum-tube computers (late 1940s), initial encounters with high-velocity missiles and space probes (1950s), nuclear power plant instrumentation (1960s; Kueser, 1965-1966), oil/gas/geothermal well-logging instrumentation (late 1970s; Sinclair, 1979; Veneruso, 1979), and special solar and Venus surface probes (early 1980s; Jurgens, 1982). Current applications include volume-limited power supplies and conversion units, industrial process tracers, and automobile engine controls. Since these applications are of limited volume, the packaging of these devices is usually expensive due to the lack of effort for further development and optimization at the package level. The high-reliability packaging schemes used today for long-service military systems do offer considerable high-temperature-operation capability (Palmer, in press).

The most widely used metal package is gold-plated Kovar (an alloy of 53 percent iron, 29 percent nickel, and 18 percent cobalt). The finish is usually a high-temperature-resistant 24-karat gold that is plated to a thickness of 100 micro-inches or greater. The underplating should be an electroless nickel plating containing 8-12 percent phosphorus that is plated to a thickness of 100-200 micro-inches (Licari and Enlow, 1988). The output leads are sealed into the Kovar package sidewalls or floor using glass-to-metal seals or ceramic feedthroughs. Metal packages consist of two configurations: plug-in or flatpack types. These metal packages can be designed with welded lids, thus assuring seal integrity at high temperatures (Palmer, in press). These packages have been evaluated to 400 °C with satisfactory results. The most common failure is due to the lack of mechanical and electrical integrity of the glass or ceramic feedthrough at high operating temperatures and after thermal cycling (Johnson, in press). As the service temperatures are increased, the electrical resistance and the hermeticity of the seal is degraded. Since the glass seal is formed at 500 °C, other types of seals need to be investigated for operating temperatures above 400 °C (Johnson, in press).

Another widely used type of package that also provides good high-temperature performance along with hermeticity is the ceramic package. The most attractive ceramic package for use in high-temperature applications is the aluminum-nitride package; however, the most often used is the less costly alumina. Ceramic packages are manufactured using a cofired tape process and have an advantage over metal packages because they can avoid the use of expensive fragile glass-to-metal seals. Ceramic packages can be designed as integral-lead packages, leadless chip carriers, and leaded chip carriers. An advantage of the integral-lead package is that the input/output leads can be spaced very closely. The packages may be sealed either by soldering or by welding. Temperature limitations for ceramic packages depend on the type of sealing method used. These packages are not suitable for high-current applications due to the resistivity of the refractory metals used as conductors. The use of packages with short tungsten conductors is also recommended since these packages will increase in resistance by about five times from room temperature to 500 °C. In addition, the insulation resistance at elevated temperature decreases by approximately 40 percent (Figure 5-1; Johnson, in press).

Today's plastic packages are not suitable for applications above 150 °C (Palmer, in press). For example, the glass transition temperature of commonly used packaging epoxies lies in the range of 130-170 °C, which limits the operating temperature range of the package. Unlike their hermetic counterpart, plastic packages subject wire bonds to extreme stresses if the package undergoes large temperature swings (e.g., -55-300 °C; Harman, in press). Silicone-gel coatings have recently been evaluated as an alternative to hermetic packages for high-reliability applications and the results have been promising. The continued use of plastic packages is highly desirable, thus the development of new higher-temperature materials is of great importance.

Besides the selection of the type of package, it is necessary to select the substrate, the component attachment method, the chip interconnection technique, and the sealing process. Understanding both material

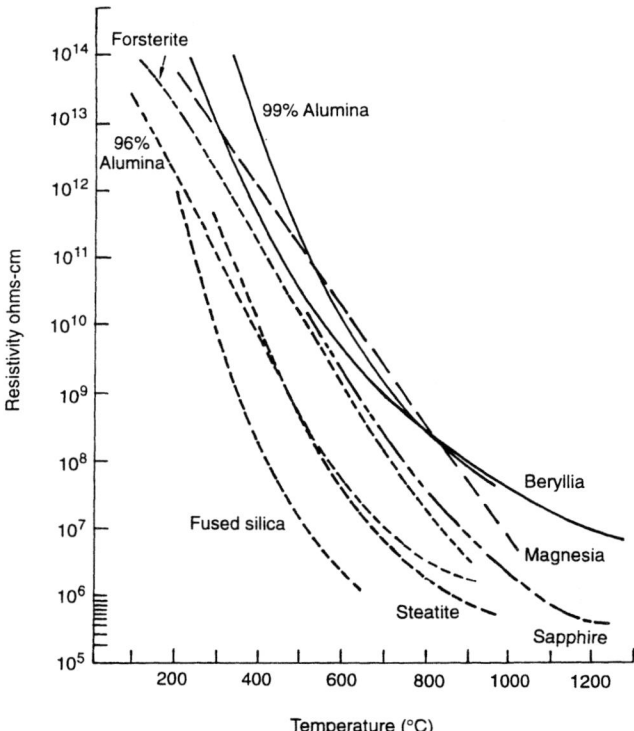

FIGURE 5-1 Decrease in insulation resistance as a function of temperature. SOURCE: Johnson (in press), © IEEE.

behavior and material interactions is of great importance. As the conventional materials are used beyond the temperature range for which they were originally developed, new failure mechanisms appear. The result is that some material systems are no longer viable. Even when the available materials are individually suitable for use at high temperatures, the devices may fail due to thermomechanical stresses or other interactions between various components of the material system. In the following section, brief descriptions of standard attachment and assembly processes are given. Supporting information on materials properties is also supplied.

SUBSTRATES

The selection of the substrate material is dictated by the combination of thermal, electrical, and mechanical design requirements for the given application. Key substrate properties for materials presently being used are given in Table 5-1. Alumina, Al_2O_3, is the most widely used ceramic substrate, and many high-temperature electronic systems have used this material without difficulty (Harman, in press). Both alumina and AlN can be used as a substrate in two different ways: they can provide the mechanical support structure for the deposition of thin-film interconnections or they can house part of the interconnection structure (cofired with molybdenum or tungsten). A significant advantage associated with the use of AlN and SiC is their high thermal conductivity and coefficient of thermal expansion, which is close to that of silicon (and almost all wide bandgap semiconductors). Diamond can also be used as a substrate material. When using such a substrate material, less stress between the substrate and chip is generated during the temperature excursions encountered in the manufacture and operation of the substrate assembly. It should be noted that SiC has a high dielectric constant and high dissipation factor limiting its use as a substrate.

The limitation of the use of AlN is the availability of chemically compatible thick-film pastes. Recently, a broad range of thick-film-paste materials for use on and in AlN substrates, including gold-based conductors, multilayer dielectric, and resistors have been developed (Tables 5-2 to 5-4; Chitale et al., 1994; Shaikh, 1994).

Diamond is a promising material to serve as a substrate for high-temperature electronics. It has the highest thermal conductivity of any material at room temperature (1,700 W/m·K). Diamond use is currently restricted by its cost and limited compatibility with thin- and thick-film metallization. However, the extremely high thermal conductivity of diamond, coupled with its electrically insulating nature and very low thermal expansion, make it a worthwhile subject for studying the manufacturing technology issues that must be resolved in order to make available 1-mm-thick diamond material in 100 mm (4 in.) or larger sizes at reasonable cost (Eden, 1994).

THICK-FILM AND THIN-FILM METALLIZATION

The desired characteristics of a metallization system are good adhesion to the substrate, low stress, good electrical conductivity, and minimal reactions at subsequent processing or assembly steps. Thick-film conductors based on noble metals have been successfully used for high-temperature applications (Bonn and Palmer, 1980; Shaikh, 1994). Typical metallurgies include tungsten, molybdenum, gold, Pt/Au, Pt/Ag, Pd/Ag, and Pt/Pd/Ag. The properties of most common metals that are considered for multilayer, cofired ceramic substrates are given in Table 5-5 (Palmer, in press).

Electrical properties of thick- or thin-film conductors differ at high temperatures. The resistivity always increases and the temperature coefficient of resistivity always changes. For example, the temperature coefficient of resistivity of thin-film gold is +0.0016/°C; for platinum it is +0.0027/°C and for tungsten it is 0.0046/°C. These resistivity changes must be taken into account for high-temperature applications. The disparity in thermal expansivities between the dielectric material (either substrate or interlevel dielectric) is about 3.7-6.8 ppm/°C, and such metals as silver (19 ppm/°C), gold (14 ppm/°C), and Ag/Pd (~19 ppm/°C) is substantial. This difference causes the metal to shrink more than the ceramic during processing and assembly or during power cycling. The cylindrical metal in the via hole begins to shrink more than the surrounding ceramic and can cause an annular void space between the via metallization and the ceramic. The most promising materials for high-temperature applications, from a mechanical point of view, are AlN substrates with cofired tungsten

TABLE 5-1 Properties of Ceramic AlN, Ceramic SiC, Glass ± Ceramics as Compared with 90 percent Alumina

Substrate Properties	Ceramic AlN	Ceramic SiC	BeO	Glass ± Ceramics	90% Al$_2$O$_3$
Thermal conductivity (W/m·K)	230	270	260	5	20
Coefficient of thermal expansion (25-400 °C) (10^{-7}/°C)	43	37	75	30-42	67
Dielectric constant at 1 MHz	8.9	42	6.7	3.9-7.8	9.4
Flexural strength (Kg/cm^2)	3,500-4,000	4,500	2,500	1,500	3,000
Thin-film metals	Ti/Pd/Au NiCr/Pd/Au	Ti/Cu	—	Cr/Cu.Au	Cr/Cu
Thick-film metals	Ag-Pd Cu	Au Ag-Pd	— —	Au.Cu.Ag-Pd Cu.Au	
Cofired metals	W	Mo	W	Au.Cu.Ag-Pd	W.Mo
Cooling capability Air (°C/W)	6	5	5	60	30
Water (°C/W)[a]	<1	<1	<1	<1	<1

[a] External cooling.

conductors. Nickel is plated on the exposed conductors and is diffusion-bonded to the base metal for enhanced adhesion. Following the nickel diffusion process, a layer of gold is deposited to prevent formation of nickel oxide and to enhance wettability during subsequent soldering or brazing processes. The final plating step is the application of heavy gold on the wiring pads to accommodate chip interconnection bonding.

The adhesion between AlN and tungsten is mechanical in nature, where joining strength is presumably provided by grain interlocking or anchoring across the interface. The microstructurally toughened interfacial adhesion also assures that the cofired multilayer substrate maintains the high thermal conductivity and mechanical strength of typical monolithic AlN (Chiao et al., 1991).

Thin-film metallizations have also been successfully used at high temperatures. The metallurgies most often used are a gold base with titanium or chromium for obtaining adequate adhesion and platinum or palladium as barrier layers. Gold is a suitable metal for this application because it has superior conductivity, protects the surface from environmental attack, and helps ensure bondability.

For diamond substrates, various thin-film conductors have been reported. For example, Au/Pt/Ti metallization was studied by Davis (1993). In this system, platinum prevents interdiffusion between titanium and gold up to temperatures of 400 °C during the annealing process. However, when annealed at 500 °C, decorated patterns are observed on the surface of the gold layer, indicating interdiffusion between the deposited metals. After annealing at 400 °C, the sheet-resistance value had increased by 40 percent. Furthermore, at 500 °C a 139 percent change in the value of sheet resistance was observed. Other examples of suitable metallurgical systems are Au/Ti-W and Au/Cr. Au/Ti-W exhibits good adhesion at temperatures up to 300 °C, but at 450 °C the adhesion degradation can be attributed to surface changes that resulted from the diffusion of Ti-W into the gold. Au/Cr appears to be more stable than Au/Ti-W, and improvements in adhesion have been observed after a 450 °C anneal.

Unless they are suitably protected, many thin films experience increases in resistance when heated in air due to their oxidation. Ultimately, they may be converted to insulating films. It has been found that the increase in

TABLE 5-2 Metallizations for AlN Substrates

	\multicolumn{7}{c}{Product Number}

	5835	D-8813	8835-1A	D-8835-1D	D-9633-D[a]	9601-N	D-9913
Metal type	Pt/Au	Gold, alloyed	Gold	Ag/Pd	Ag/Pd	Ag/Pd	Silver
Fired thickness, μm	12-15	8-10	8-13	7-10	9-12	12-15	9-13
Resistivity, mΩ/□	<60	<4	<7	<4	<35	<3	<2
Adhesion[b], initial, Kg	>4.5	>5.0	>2.5	>2.5	>6.0	>5.0	>5.0
Adhesion, 200 hrs at 150 °C, Kg					>3.0	>3.0	>3.0
Thermosonic 25μm gold wire, g		>14	>12	>14	>6	>6	
Ultrasonic 250 μm aluminum wire, g			>400	>250	>350	>250	

[a] This Ag/Pd conductor has the highest initial adhesion. Unfortunately it is incompatible with all other ESL-thick-film materials for AlN.
[b] Adhesion, 90° pull, 2 mm x 2 mm pads.

resistance of a film is considerably greater than can be accounted for on the basis of only reduced thickness of the conductive portion of the film resulting from oxidation of the surface. This is due to the fact that oxidation occurs along the grain boundaries of the film as a result of either oxygen diffusing in from the surface or from oxygen trapped internally that precipitates at the grain boundaries and forms an insulating phase.

COMPONENT ATTACHMENT

The primary method of securing the device to the substrate (die attach) is by diffusional reaction of the backside (usually metallized) with a substrate gold metallization. This joint is designed to elastically absorb the thermomechanical stress that accumulates between the chip and the ceramic substrate, conferring excellent fatigue resistance. There are three types of die-attach materials that are most suitable for application in high-temperature electronic assembly (i.e., hard solders, glass-based die-attach materials, and the recently developed amalgams). Hard solders provide electrically conductive paths and have high thermal conductivity but are difficult to rework. They also require high-temperature processing conditions (depending on the alloy). They can be rigid and brittle, causing cracking of a large die. Tests with Au/Si and Au/Ge have satisfactorily demonstrated small chip attachment (less than 9 mm on a side; Draper and Palmer, 1979). The disadvantage of using hard solders stems primarily from their lack of plastic flow, which leads to high stress in the devices because of the thermal expansion mismatch between the die and the substrate.

C.A. MacKay (1991) developed a new method of making a bonding amalgam that has potential for application in high-temperature electronics as die attach, hermetic sealing, and heat-sink die attachment. An amalgam is defined as a nonequilibrium, mechanically alloyed material formed between a liquid metal and a powder. In general, amalgams have low processing temperatures. Nevertheless, they still yield materials with thermal stabilities between 250 °C and 600 °C.

The use of molytabs might provide a solution for reducing the high stresses that the hard solders impose on the system. In this technique, the die is preattached to gold-plated molybdenum tabs (molytabs) and then solder-attached to the tape automated bonded (TABed) die from the substrate bonding pads. The molytab, with a coefficient of thermal expansion of approximately 5 ppm/°C, forms a buffer between the device and the substrate.

TABLE 5-3 Dielectrics for AlN Substrates

Product designations	ESL# D-4907, ESL# D-4907-A
Thermal expansion coefficient	44 x 10^{-7}/°C
Dielectric constant (@ 1 MHz)	5 - 7
Dissipation factor (@ 1 MHz)	0.5 - 0.8%
Insulation resistance (100 VDC, 50 μm thickness)	>10^{11} Ω
Breakdown voltage (50 μm thickness)	>2 kV
Adhesion to AlN	Exceeds substrate fracture strength
Compatibility	Most ESL Au and Ag conductors ESL# R-300 resistor series (when printed on the dielectric)

Silver-glass-paste adhesives have also found an application in the assembly of high-temperature electronic devices. These materials are reflowed at temperatures of 400-450 °C. The glass recrystallizes during the reflow cycle and can subsequently be exposed to temperatures in excess of 400 °C (Johnson, in press). These materials offer the possibility of the formation of void-free, die-bond interfaces with excellent thermal stability. The tendency for silver migration and the effect of thermal cycling on the long-term adhesion of this system over extended temperature ranges has not yet been studied, however.

Polymer die attach is normally not practical for use at temperatures above 250 °C where a nonconductive attachment is needed. However, silicone/polyimide adhesives can be made to operate effectively up to 400 °C. These materials offer the lowest cost when compared with the gold-based hard solders. The use of organic adhesives also supposedly lowers the thermal stress in devices. However, the use of organic adhesives in high-temperature electronic packages has been limited because of outgassing and poor thermal stability (especially when filled with silver).

INTERCONNECTION

The typical interconnection methods used between chips and substrates are wire bonding, tape automated bonding, and flip-chip. Depending on the required operating temperature, the current interconnection methods may be limited by their material and mechanical properties for elevated temperature applications. Wires and bonds in a mono-metallic situation have been evaluated for temperatures as high as 500 °C for up to thousands of hours and appear to be relatively trouble-free for either Au-Au or Al-Al (Harman, in press). The application of parallel-gap welding for bonding of 5-mil annealed platinum wire to a 5,000 Å platinum pad (with an underlying base of titanium, molybdenum, and tungsten) sputter-deposited on a sapphire substrate has been successfully developed for a high-temperature interconnect (800-1000 °C) for a thermionic integrated circuit (a vacuum integrated circuit technology; Fendrock and Hong, 1990).

Problems arise when dissimilar metals are joined. For example, in case of Au-Al intermetallic compounds, Kirkendall voids can form, weakening the interface. Another area of concern is the thermally induced change that can take place within some metallic materials themselves (Harman, in press). Two examples are reconstruction and creep in solders and crystallographic and electromigration changes in thin aluminum films and wire. Solders may be reliably operated at temperatures close to their melting point. Either substitute materials must be developed (for flip-chip) or all connections including package leads to the board must be welded either by using thermocompression or thermosonic methods (Johnson, in press). For example, the gold bumps in TAB provide connections between the aluminum pads (separated by a barrier metal to prevent diffusion and intermetallic formation at Al/Au interface) and the copper tape. These inner lead connections are made using a thermocompression technique. For outer lead bonding some type of welding is recommended. Exposed copper metal on tape should be completely plated with gold to avoid copper oxidation. While this technology shows great promise for high-temperature electronic applications,

TABLE 5-4 Summary of Properties of Metallizations for AlN

	Wirebondable Conductors			
	FX 30-010	FX-30-011	FX-30-022	FX-30-023
Metallurgy	Gold	Au/Pd	Gold	Au/Pd
Application	Multilayer	Multilayer	Single layer	Single layer
Wire type	Gold	Silver	Gold	Al
Resistivity (mΩ/□ at 1 mil)	2.5-3.5	4.0-5.5	2.5-3.0	4-5
Wirebond strength (g)	9-11	11-13	10-12	12-14
Thermal conductivity $W/m^{-1}K^{-1}$	180	180	180	180

	Solderable Conductors	
	FX 34-100	FX 31-012
Metallurgy	Ag/Pd	Pt/Au
Resistivity (mΩ/□ at 1 mil)	13-16	16-181
Solderability (seconds to 100% coverage in 60/40 Sn/Pb)	1	1
Solder leach resistance (1 s dips to 80% retained)	25-30	20
90° peel adhesion (1 lb on 80 mil x 80 mil pad) Initial	10-14	8-13
Aged (150 °C, 48 hrs)	8-10	2

bumped die and tapes are not readily available in the United States.

SECOND-LEVEL PACKAGING

Traditional packaging solutions for high-temperature applications have usually depended on inorganic substrates (e.g., alumina) and refractory metal conductors (e.g., tungsten and molybdenum). Since these materials require processing temperatures in excess of 1000 °C, their performance at high operating temperatures is inherent. The connections to these refractory conductors must also tolerate the operating environment and are therefore formed by welding, brazing, or high-temperature soldering (i.e., metal-loaded glass frit solder). Selection of materials for these connections must accommodate the increased diffusion caused by long-term, high-temperature operation.

Recent advances in organic dielectrics have allowed some application of organic, printed circuit boards in high-temperature environments, primarily for reduced cost. Polyimide printed circuit boards patterned with nickel, nickel-plated copper, or nickel-plated tin are common combinations. The bondability of nickel for welding or soldering is often improved by the selective deposition of gold over the bond sites.

Some applications, however, require the use of large very-large-scale-integrated (VLSI) devices that would be unreliable if attached to substrates with significantly

TABLE 5-5 Typical Cofired Metals

Metal	Melting Point (°C)	Electrical Resistivity (10^{-8} Ω·m)	Thermal Expansion (10^{-7}/°C)	Thermal Conductivity (W/m·K)
Silver	960	1.6	197	418
Gold	1063	2.2	142	297
Copper	1083	1.7	170	397
Palladium	1552	10.8	110	71
Platinum	1774	10.6	90	71
Molybdenum	2625	5.2	50	146
Tungsten	3415	5.5	45	201

different coefficients of thermal expansion. In these applications, silicon, or other ceramics, is the substrate material of choice. Conductors of patterned tungsten are used, although aluminum can be used if sufficiently protected with a high-temperature dielectric such as phospho-silicate glass or silicon nitride.

SUMMARY

Although high-temperature wide bandgap devices are being developed, they can only be successfully implemented into commercial applications if the packaging technology is developed in parallel to provide a cost and performance advantage. Novel approaches should be examined based on the existing knowledge of advanced packaging technologies, especially multichip modules. All of the integrated circuit wide bandgap devices for high-temperature electronics that are being developed are of a lower density of integration than existing complementary metal-oxide-semiconductor (CMOS), silicon-on-insulator (SOI), or even GaAs devices. Although existing packaging technologies can support the operation of high-temperature electronic integrated circuits, SiC, AlN, and diamond technology for packaging applications should be examined since the use of these materials will minimize the thermal mismatch and reduce any resulting stresses. The utilization of diamond and AlN as dielectric materials for multilayer interconnects deposited either on cofired AlN or on blank AlN, SiC, and diamond substrates should also be a part of the investigation. The use of SiC-encapsulated semiconductor devices (as demonstrated by Dow Corning) should also be considered. This technology ensures a hermetic die, which is very important for the devices working at elevated temperatures because chemical reaction rates are much higher at these temperatures. SiC encapsulation should be examined in conjunction with area array connections. Area array connections, such as tungsten bumps overplated with nickel and gold or copper bumps, are more suitable for high-temperature electronics than the traditional PbSn flip-chip attachment method. Other metallurgical structures should also be examined. Although there is no immediate need for the implementation of flip-chip or multichip module technology in high-temperature electronic applications, the benefits of developing these technologies could substantially affect the cost and performance of high-temperature electronics. For example, when using high-temperature electronics in conjunction with multichip module technology, the reliability of the system is improved due to a reduction in the number of metallurgical connections. In multichip module packages, the signal travels directly from chip to chip, all within the protection of a hermetic environment (Figure 5-2; Pedder, 1988).

As an interim solution, the following packaging schemes that are based on existing packaging technologies can be used in the temperature ranges specified below:

- For temperatures <200 °C
  —nonhermetic with silicones
  —metal (welding, AuSn, gold diffusion)
  —ceramic (welding, AuSn, glass, gold diffusion);
- For temperatures <300 °C
  —metal (welding, AuSn, gold diffusion)
  —ceramic (welding, AuSn, glass, gold diffusion);

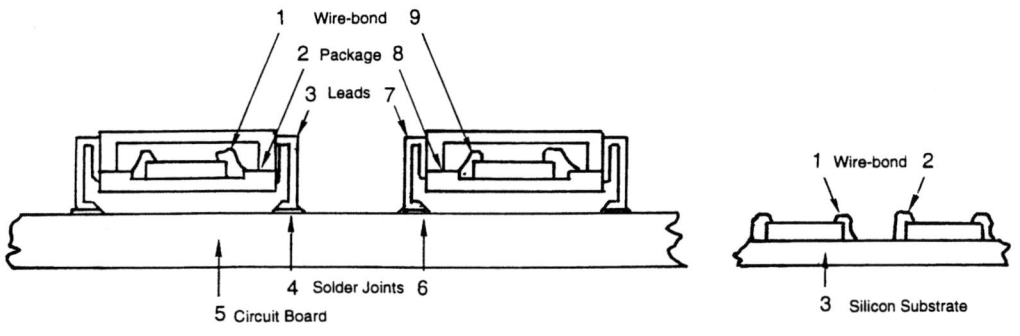

FIGURE 5-2 Reduction from nine to three electrical path segments between two integrated circuits with multichip module technology. SOURCE: Pedder (1988).

- For temperatures <400 °C
  —metal (welding, gold diffusion)
  —ceramic (welding, glass, gold diffusion); and
- For temperatures <500 °C
  —ceramic (welding, gold diffusion)
  —sealed chip technology: Si_3N_4, SiO_2, diamond.

While materials and processes exist to meet the packaging challenges of high-temperature electronics whenever the market can match its needs with the supplier, better packaging solutions can be developed for high-temperature electronic applications when more complex and powerful systems are required. The use of existing high-density multichip module designs can leverage the development of high-temperature electronic systems by careful selection of the material systems used.

There is not an upper temperature that should be feared by packaging engineers any more than by the device physicist (Johnson, in press). Since high-temperature electronic devices and systems may undergo large temperature excursions depending on their applications (e.g., jet engines and spacecraft), the material and electrical properties will change accordingly (Sampson and Mattox, 1991). This change will be accompanied by the structural stresses on the chip and its package that are caused by the different thermal coefficients of expansion. Therefore, there is a need for an electrical and mechanical simulation for the proposed circuit. Nevertheless, basic material properties (for example, intermetallic compounds) must be examined over the wide temperature range that may be encountered in high-temperature electronics on both the short-term and long-term high-temperature behavior of the materials systems used in assembly and packaging (Harman, in press).

6

Device Testing for High-Temperature Electronic Materials

This chapter concentrates on devices made from silicon and silicon carbide, as these are the two materials systems for which the most data are available. The data are encouraging and suggest that electronic systems based on either of these two materials will operate successfully at elevated temperatures.

There are three areas of testing that are discussed in this chapter: (1) short-term, constant-temperature tests, (2) constant-temperature life tests, and (3) thermal-cycling tests. The short-term and constant-temperature tests are encouraging. There is no known thermal-cycling data, however, and more research is required. A few comments at the conclusion of the chapter are devoted to the subjects of packaging and future testing

SHORT-TERM CONSTANT-TEMPERATURE TESTS

Short-term constant-temperature tests are the first type done on each device intended for use at elevated temperatures and thus the one test for which there is the most data. In a typical test of this kind, a device or integrated circuit is placed in a test fixture that can be varied in temperature. Often, the device under test is also in a special, controlled environment, such as dry nitrogen or vacuum.

Figures 6-1 and 6-2 are examples of two such tests. A representative set of conditions, with some associated comments, is given in Table 6-1. Figure 6-1 illustrates the variation of threshold voltage with temperature for state-of-the-art silicon MOSFETs. Both n- and p-channel devices change from enhancement mode to depletion mode at about 350 °C. Figure 6-2 illustrates the drain characteristics of a SiC MOSFET at 650 °C. The

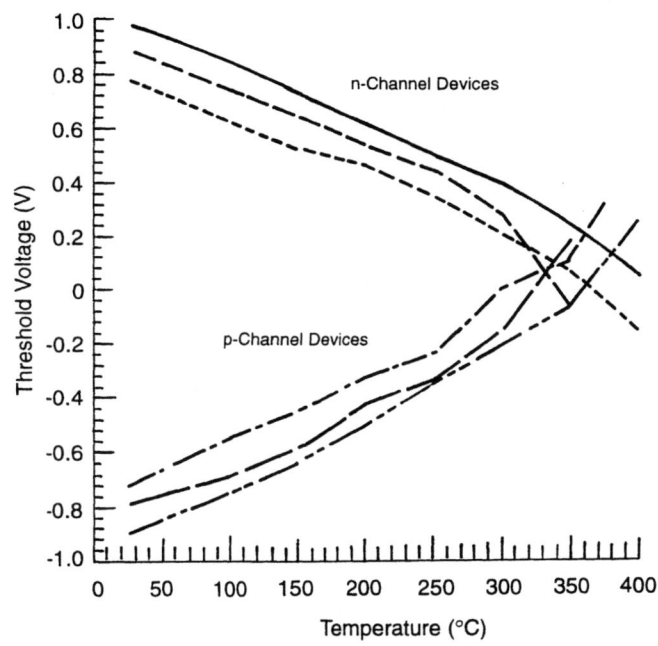

FIGURE 6-1 Variations in threshold voltage for p- and n-type silicon MOSFETs with temperature. SOURCE: Grzybowski and Tyson (1993).

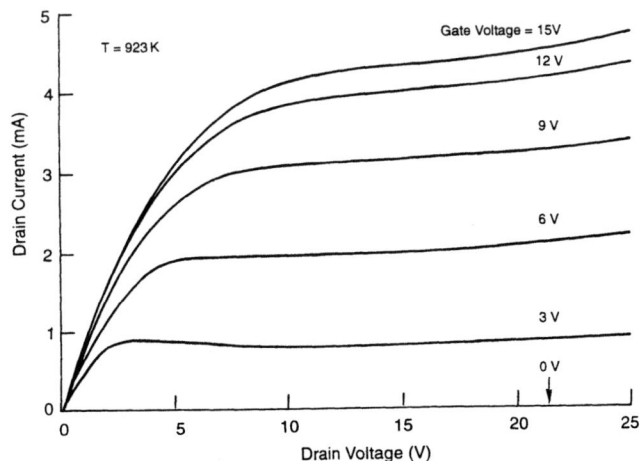

FIGURE 6-2 Drain characteristics of a SiC inversion-mode MOSFET at 650 °C. SOURCE: Palmour et al. (1991).

TABLE 6-1 Short-Term Constant-Temperature Tests

Device Type	Temp. (°C)	Comments	Substrate	Attachment Method	Reference
SiC transistors	25-650, unpacked	Gain increases with temp.	BN plate	Not reported	Palmour et al., 1991
SiC transistors	25-350, packaged	Prototypes available	Not reported	Not attached (wafer probe)	Palmour, 1993
SiC diode	25-550	Very low leakage currents	Not reported	Not reported	Ghezzo et al., 1992
SiC pressure sensors	25-400	Piezo-resistive bridge devices	None	Not reported	Shor et al., 1994
SiC amplifiers	25-350 (as a hybrid IC)	Small gain change with temp.	Alumina or glass ceramic	Braze	Tomana et al., 1993
Silicon transistors	-65-450 (both FETs and bipolar)	Gain decreases with temp.	Alumina	Silver glass	Grzybowski, 1991; Grzybowski and Tyson, 1993

characteristics are normal for MOSFETs, with low leakage and relatively constant gain.

These results clearly indicate that devices in both materials systems have been made that operate at temperatures well above the current limits of fielded electronic systems, typically about 125° C. The results are also consistent with the fundamental limits of device operation, based on the characteristics of each material of energy gap, effective mass, minority carrier lifetime, and leakage currents.

CONSTANT-TEMPERATURE LIFE TESTS

Constant-temperature life tests are also encouraging (Table 6-2), although much work remains to be done in this area. The times on tests at the temperatures indicated are too small to be conclusive, however. Effects such as metal electromigration, impurity diffusion in the semiconductor, and phase changes occurring in contact regions, have not yet been adequately tested. A possible exception is the last result, which was based on silicon integrated circuits with conventional metallization and feature sizes that are large (about 7.5 micrometers) by current standards. This result suggests that silicon-based circuits with conventional metallization may be useful to temperatures well above the current limit of 125 °C.

THERMAL-CYCLING TESTS

Thermal-cycling tests logically follow after the short-term and constant-temperature tests. Thermal-cycling tests are absolutely essential for evaluating suitability of devices intended for use at elevated temperatures. It is well known, and is a major limitation in conventional temperature electronics, that thermal cycling results in failures that are not discovered by constant-temperature tests. One class of these failures is associated with the work-hardening of parts of the device, its metallization, the package attachment, and the package itself. In flip-chip technology, for example, the work-hardening of the solder bumps that are used to hold the device in place and form the interconnects is a well-known limitation. In fact, a substantial amount of development has been devoted to minimizing this problem. For elevated-temperature electronics, these mechanisms will almost certainly be more of a limitation, since the temperature swings will be larger.

TABLE 6-2 Constant-Temperature Life Tests

Device Type	Time (hrs)	Temp. (°C)	Substrate	Attachment Method	Reference
SiC diode rectifiers	1,000	350	Hermetic glass packages	Not reported	Edmond et al., 1991
Silicon MOSFETS	1,000	200	Alumina	AlSi eutectic	Palmer and Heckman, 1978
Silicon bipolar quad op-amp	500	300	Not reported	Not reported	Beasom and Patterson, 1982
Silicon-based ring oscillators	4,000	250	Not reported	Not reported	Migitaka and Kurachi, 1994

These limitations may demand new and innovative techniques for device attachment and connection, such as the use of graded attachment techniques or compliant attachment methods, in which the differences in thermal expansion are taken up by a relatively flexible part of the attachment. The use of different attachment materials may also be useful. For example, the use of amalgams has been explored on a preliminary basis. The use of these materials would allow one to make the rigid attachment at a temperature that is intermediate between the extremes, thus minimizing the stress that occurs at either extreme.

FUTURE REQUIREMENTS FOR HIGH-TEMPERATURE TESTING

The testing of devices, circuits, and systems intended for high-temperature operation is more difficult than testing for lower-temperature situations. For these lower-temperature applications, the concepts of step-stress testing and accelerated aging are established. In these two approaches, the device under test is subjected to increasingly higher temperatures and the failure rates noted. In a well-behaved test, the resulting failure rates will allow the calculation of an activation energy, which will in turn allow the prediction of failure rates at lower temperatures.

The situation, however, is less clear for higher-temperature devices. For example, the mere testing at higher temperatures is a challenge, with a lack of equipment available for these higher-temperature tests. Also, a key assumption for accelerated testing is that the mechanisms of failure are the same for the accelerated test as for the application. This assumption has not been shown to be valid, and may not be true for many tests. For example, it is likely that the material used to mount the devices to the substrates would melt in accelerated testing, thus introducing a new failure mechanism and invalidating the test.

A possible solution to this problem is the continuous-variables testing method. This method involves the measurement of parameters with very high resolution. In these tests, the device conditions are similar to those of the intended application. By careful monitoring of the device parameters, failure mechanisms can be detected at early stages. Use of this technique eliminates the problems associated with step-stress and accelerated testing, in which new failure modes may be introduced.

7

Conclusions and Recommendations

There are great opportunities for wide bandgap semiconductors to improve the performance of many nonelectronic technologies. Major benefits to system architecture would result if cooling systems for components could be eliminated without compromising system performance (e.g., power, efficiency, speed). The existence of *commercially available* high-temperature semiconductor devices would provide significant benefits in such areas as:

- sensors and controls for automobiles and aircraft;
- high-power switching devices for the electric power industry, electric vehicles, etc.; and
- control electronics for the nuclear power industry.

With the possible exception of LEDs, however, present commercial demand for wide bandgap semiconductor materials is limited. While there are few pressing applications that cannot be achieved without wide bandgap materials, the vast array of applications, and hence the value, will only be realized once these materials have evolved to such an extent that off-the-shelf devices are available.

This chapter is divided into two sections. The first section presents general conclusions and recommendations about future research priorities to accelerate the acceptance of high-temperature semiconductor materials. This section discusses the temperature ranges for the different materials to be used, the competitiveness of U.S. research versus foreign competition, the systems in which high-temperature electronic materials should initially be introduced, and the government/industry/university collaborations required to forward the development of high-temperature semiconductor materials. The second section discusses the barriers to the successful development, manufacture, packaging, and integration of wide bandgap materials into existing systems and presents the key research and development priorities to overcome these barriers.

GENERAL CONCLUSIONS AND RECOMMENDATIONS

Temperature Ranges

Silicon and silicon-on-insulator (SOI) electronics may be sufficient for some applications for temperatures up to 300 °C. Such applications include digital logic, some memory technologies, and some derated analog and power applications. Silicon-based technology will not be sufficient for many applications operating in the 200-300 °C range, however, such as power-conditioning devices in higher-temperature control systems. These devices will have to be produced from another material system. *Devices based on SiC are well positioned to meet this need, particularly n-channel enhancement-mode MOSFETs. However, significant technological barriers, such as micropipes, oxide quality, contacts, metallization, packaging, and reliability evaluation still need to be further addressed.*

As a result of fundamental limitations, silicon-based technologies will not be useful at temperatures above 300 °C. Other materials must be used for these temperature ranges, but the choices are somewhat less clear. Technology based on GaAs might be used for systems operating up to 400 °C. Just working at elevated temperatures is not the only concern, however. It is also essential that the devices reliably function over a wide range from very cold (i.e., -20 °C) to very hot (i.e., 400 °C). *Based on the evidence presented in this report, devices based on n-type SiC are the only type that currently appear to meet the temperature-range and reliability requirements, but additional development is needed.* Eventually, high-temperature electronic technology could be developed for reliable operation even for temperatures above 600 °C.

U.S. Competitiveness

As described in the Preface, considerable international resources are currently being devoted to developing electronic technologies either tailored for or supportive of high-temperature operation. The United States is focusing most of its efforts on high-temperature applications and currently has a slight lead in SiC research.

Europe appears to be increasing its effort in wide bandgap materials, especially for power electronics. This research area is synergistic with high-temperature applications because the generation of internal heat is a limiting factor in power devices and can be mitigated by larger bandgap and higher thermal conductivity materials. The dedication of European resources to this area is seen in the founding of the collaborative organization HITEN, which was established in 1992 to coordinate nascent European efforts in high-temperature electronics.

Japan is emphasizing the use of wide bandgap materials for opto-electronics and leads in the use of nitrides for light sources. Japan is also becoming interested in power and high-temperature applications. Unfortunately, the closed nature of Japanese industry made it difficult for the committee to determine the true level of interest in wide bandgap materials research. The increased interest in high-power, high-temperature applications is evident in Japan's annual domestic SiC conference, however. The Third Domestic (Japan) SiC Conference convened in Osaka on October 27-28, 1994, with approximately 160 experts in attendance. Contrary to Japan's previous two conferences, there was a greater emphasis at the Osaka conference on high-power, high-temperature applications than on LEDs.

The Commonwealth of Independent States had a number of major programs in SiC development, but the current financial difficulties of most of the Commonwealth's institutions are preventing many laboratories from continuing their research. There is a wealth of expertise and information available for leveraging by other countries, however. For instance, the European Community is planning on supporting a SiC growth effort in St. Petersburg (Y.M. Tairov and V.E. Chelnekov, personal communication, 1994).

The committee believes that the U.S. wide bandgap materials research community is currently very competitive in the international research community. *To remain competitive in the international research community, the committee recommends that demonstration technologies be pursued to motivate further research and increase interest in high-temperature semiconductor applications.*

Demonstration Technologies

To increase interest and motivate further research in wide bandgap materials, a realistic, inspiring application focus must be found that can make system designers aware of the benefits of high-temperature electronics. A wide bandgap transistor that operates at 150 °C will not drive the technology because it will be in direct competition with the more economically efficient silicon technologies. The demonstration technologies must be *system* circuits (i.e., not an *individual* device) that can be inserted into essentially nonelectronic systems (e.g., turbine engine, nuclear reactor, chemical refinery, or metallurgical mill) with the goal of measurably increasing system performance.

As discussed in Chapter 1, the committee believes that there eventually will be a niche market for semiconductors with temperature capabilities higher than that of silicon, and that this market will be sufficiently large to justify the cost of development. However, this belief is tempered by the recognition that because such electronics will be used in new ways there is little immediate demand. The market will grow only in synergy with the availability of components. This suggests that development of high-temperature electronics not be undertaken in isolation. Instead, such development can and should be leveraged from development of other technologies with more immediate applications, thus reducing the costs and the risks of both. Three suitable application areas are high-power electronics, nuclear reactor electronics, and opto-electronics.

Power switching devices, for example, would be a good demonstration technology for high-temperature semiconductor materials. High-voltage, high-power electronics, while not necessarily used as high-temperature devices, nevertheless need wide bandgap semiconductors because of their superior breakdown voltages and high thermal conductivities. There is already considerable research being pursued in this area because (1) improved high-power switching devices could save an estimated $6 billion in the cost of construction of additional transmission lines; and (2) the smoother, more efficient use of the transmission system would reduce the need for new generating capacity, which the Electric Power Research Institute estimates would be a savings of $50 billion in North America alone over the next 25 years (Spitznagel, 1994).

The pursuit of demonstration technologies would not only increase interest in wide bandgap materials, it would also provide significant testbeds for the application of the technology and enhance our understanding of the *generic* technologies required to further high-temperature device operation (e.g., materials etching and implantation; degradation modes of metallic gates, contacts, and interconnects at high temperatures; packaging behavior at high temperatures; and accelerated-testing and reliability-testing methodologies to ensure proper functioning). *The ability to grow a reasonably defect-free material is not the only requirement for the realization of a successful technology. The development of demonstration technologies would also help identify other factors that must be resolved for high-temperature electronics to be incorporated into existing systems.*

Funding Strategy

The need for new development funds for demonstration technologies and future wide bandgap materials is not necessary in the committee's opinion. Government funding currently exists for long-range research in wide bandgap materials, although additional funding would certainly allow more options to be evaluated within a shorter period of time. Industry has also demonstrated a willingness to commercialize new developments if the projected payback to their investments can occur within the short term (NRC, 1993). *The committee believes that the high-temperature research community should leverage the research funding for wide bandgap materials that is currently being provided by the high-power and optics markets, where no viable alternatives to wide bandgap materials currently exist.* Building on the funding for other areas dependant on wide bandgap materials reduces the need for potential users of high-temperature devices to fund the required materials development exclusively and, thus, may render it cost effective.

The committee recommends the following strategy for the development of wide bandgap materials:

- develop precompetitive alliances and integrated programs (national laboratories, universities, and industries) for coordinating research, technical skills, and capabilities to expedite research in the most efficient manner;
- direct research at a technology demonstrator that has definite applications (i.e., is a product) and addresses the usually neglected areas of packaging, assembly, testing, and reliability (e.g., high-power switches; integrated motor control; power phase shifter);
- concurrently develop materials, design, testing, and packaging; and
- build and test the demonstration component on a cost-share basis that encourages teaming, ensures adequate funds, and requires periodic deliveries.

The committee believes that the founding of a newsletter that provides a summary of published worldwide developments in high-temperature semiconductor research would assist the establishment, development, and maintenance of (1) a fundamental long-term materials effort, (2) an infrastructure within the industry, (3) a group to monitor international development, and (4) a U.S. information group for highlighting advances.

MATERIALS-SPECIFIC CONCLUSIONS AND RECOMMENDATIONS

The first three parts of this section concentrate on the major wide bandgap materials discussed in this report: SiC, nitrides, and diamond. The final part of this section concerns the generic problems in packaging that will affect the production of all high-temperature electronic devices.

Silicon Carbide

SiC is an indirect bandgap semiconductor and has enjoyed the longest history and greatest development with regard to both materials growth and device realization. As such, SiC is currently the most advanced of the wide bandgap semiconductor materials and in the best position for near-term commercial application. Its main application will be in high-power, high-temperature, high-frequency, and high-radiation environments. It will not be suitable for blue lasers or ultraviolet light emitters, however, except as a potential substrate material. The specific technical issues for SiC that require further research are summarized in the box, *Technical Issues for SiC*. The three key research efforts for the development of commercially viable SiC devices are

- *Wafer production:* The 1- and 2-inch SiC wafers now in production are rapidly approaching *device quality* where they might be used for commercial production of devices and circuits with acceptable yield. It could be argued that such small wafers are entirely sufficient for what will be a relatively small market (compared with silicon) with a very high-price premium, and therefore an early investment in larger wafers is not justified. However, the entire commercial infrastructure for electronics manufacture is based on a wafer size of at least 3 inches, and preferably 4 inches, as a minimum. Reconstructing a small-wafer infrastructure that became obsolete over 30 years ago will be both an expense and an obstacle to the introduction of commercial SiC electronics. The committee believes that the development of larger SiC wafers is viewed as the more cost-effective approach to commercial development.
- *Film growth:* Chemical vapor deposition, molecular-beam epitaxy, and other film-growth technologies and chemistries require refinement to produce epitaxial films with n- and p-type doping ranges from 10^{13} to 10^{20} cm^{-3} for nitrogen, aluminum, boron, gallium, transition metals, and rare earth elements.
- *Manufacturing processes:* Lower-cost device-production methods are required to make the manufacture of SiC devices more competitive with the silicon technologies.

Technical Issues for SiC

Further improvement of crystal perfection (boule and CVD growth)
   Eliminate micropipes
   Reduce defect density
   Reduce background impurities
   Improve surface morphology
Further improvement of doping (boule and CVD growth)
   New n- and p-type dopants
   New mid-gap impurities for semi-insulating substrates
   Introduce rare earth elements in growth
Improve processing (boule and CVD growth)
   Improve oxides/passivation
   Find alternative insulators (nitrides)
   Reduce contact resistivity for p-type material
   Develop high-temperature n- and p-type contacts
   High-temperature packaging
Improve understanding of basic properties and knowledge of design parameters

---

Technical Issues for Nitrides

Substrate development
   Nitride substrates for CVD homo-epitaxy
   High thermal conductivity, quasi-lattice matching substrates (both high electrical conductivity and semi-insulating)
Further improvement of crystal perfection and doping (CVD growth)
   Reduce defect density and background impurities
   Better control of n- and p-type doping
   New technologies for epitaxial growth
   Improve surface morphology
Improve processing (CVD growth)
   Ohmic contacts
   Low contact resistance
   High-temperature contacts
   High-temperature packaging
Improve understanding of basic properties and knowledge of design parameters

---

Nitrides

Interest in the direct bandgap nitride materials (i.e., GaN, AlN, AlGaN, and InGaN) has dramatically increased recently because of their optical properties. The materials show great promise and are likely to dominate the visible and ultraviolet opto-electronics market. Nichia's recent bright blue LEDs have already stimulated increased industrial effort (e.g., Hewlett Packard, Spectra

Diamond

Diamond is a well-understood material, but its use for active electronic device applications is not feasible at this time because of the difficulties associated with its economical growth and doping. While diamond transistors have been designed, fabricated, and tested, their performance is also orders of magnitude less than that which is expected from the electrical properties intrinsic to diamond. The poor performance is thought to result from excessive nitrogen impurities and from as yet not fully explained surface-depletion effects. The current prognosis for diamond is primarily as a protective coating, a thermal management film, and a material for electron-emitting cathodes. The specific technical issues for diamond research are summarized in the box, *Technical Issues for Diamond*.

Technical Issues for Diamond

Improvement of growth, crystal perfection, and growth
   Reduce and control impurities of bulk synthetic diamonds
   Produce large-area (hetero-epitaxy) single-crystal films of diamond on nondiamond substrates at reasonable cost
   Synthetically produce larger bulk diamond at reasonable cost
   Improve n- and p-type doping
   Improve implantation for doping
Improve processing
   Ohmic contacts
   Low contact resistance
   High-temperature contacts
   Hydrogen passivation
Improved understanding
   Dopant diffusion
   Knowledge of design parameters

Diode Laboratories, Xerox PARC) in materials growth, contact metallurgy and reliability, and device reliability and testing, although the materials have defect densities of greater than $10^{10}/cm^2$ and the mechanism of photo emission is currently unknown. Heterojunctions in the nitrides also hold promise for higher-speed devices compared with SiC. Their applicability for power development and high-frequency devices is unproven at this time, and the technologies for wafer production, doping, and etching are currently less developed than SiC and require more longer-term research before they will be competitive with other electronic materials. However, as development of photonic applications for wide bandgap materials progresses, the opto-electronic market may provide an effective way to leverage the development of these materials for high-temperature device applications. The specific technical issues for nitrides research is summarized in the box, *Technical Issues for Nitrides*. The committee identified the following three research efforts as being key to the development of nitride devices:

- *Compatible substrates:* Better-matched substrates are required for nitride wafer production to be commercially tenable.
- *Wafer production:* Growth of quasi-crystalline films of GaN, AlGaN, and AlN should be pursued on substrates such as SiC to gain thermal advantages.
- *Doping:* Methods for both n- and p-type doping of Group III nitrides are required.

Packaging

Much more research is required in the area of high-temperature packaging. For high-temperature electronics to be commercially viable and provide true performance advantages, interconnection and packaging technologies are required that can reliably operate at temperatures up to 600 °C for 10^4 hours. To attain these goals, innovative packaging techniques will be required. The specific technical issues for packaging research are summarized in the box, *Technical Issues for Packaging*. The three key research efforts for the development of high-temperature packages are

- *Metallization:* Contacts are required in the 10^{-6} to 10^{-7} Ω/cm^2 range that have long-term durability at temperatures up to 600 °C. Greater understanding is needed of the long-term effects

---

Technical Issues for Packaging

Improve reliability of high-temperature contacts
   Improve metallization
Improve device development tools
   Improve process-control tools
   Improve polishing, cutting, mounting, and etching methods
   Develop reliability and aging tests
   Develop computer-aided design tools

of high temperatures on contact and interconnect metallurgy, degradation and failure modes, reliability, and interfaces.
- *Device reliability and aging testing:* Existing methods of accelerated, environmental life testing of packages must be adapted for high-temperature applications to ensure the accurate assessment of device reliability and aging.
- *Computer-aided design tools:* Computer-aided design tools are required that incorporate electrical and mechanical simulation of high-temperature electronic systems.

References

Acheson, E.G. 1893. Journal of the Franklin Institute. September:194.

Adesida, I., A. Mahajan, E. Andideh, M.A. Khan, D.T. Olsen, and J.N. Kuznia. 1993. Reactive ion etching of gallium nitride in silicon tetrachloride plasmas. Applied Physics Letters 63(20): 2777-2779.

Alok, D., M. Bhatnagar, H. Nakanishi, B.J. Baliga, Y. Chang, and R.F. Davis. 1993. 3C-monocrystalline SiC reactive ion etching using SF_6/O_2. Pp. 585-587 in Proceedings of the 5th International Conference on Silicon-Carbide and Related Materials. M.G. Spencer, R.P. Devaty, J.A. Edmond, M. Asif Khan, R. Kaplan, and M. Rahman, eds. Institute of Physics Conference Series #137. Bristol, England: Institute of Physics Publishing.

Amano, H., M. Kito, K. Hiramatsu, and I. Akasaki. 1989. P-type conduction in Mg-doped GaN treated with low-energy electron beam irradiation. Japanese Journal of Applied Physics 28(12.2):L2112-L2114.

Anthony, T. 1994. Personal Communication to W.J. Choyke.

Arbab, A., A. Spetz, Q. ul Wahab, M. Willand, and I. Lundstrom. 1993. Chemical sensors for high temperatures based on silicon carbide. Sensors and Materials 4(4):173-185.

Baliga, B.J. 1982. Semiconductors for high-voltage, vertical channel FET's. Journal of Applied Physics 53:1759-1764.

Balooch, M., and D.R. Olander. 1992. Etching of silicon carbide by chlorine. Surface Science 261(1-3):321-324.

Beasom, J.D., and R.B. Patterson. 1982. Process characteristics and design methods for a 300 °C quad operational amplifier. IEEE Transactions on Industrial Electronics IE-29(2):112-117.

Benjamin, M.C., C. Wang, R.F. Davis, and R.J. Nemanich. 1994. Observation of Negative Electron Affinity of Heteroepitaxial AlN on 6H SiC. Paper D1.12. Paper presented at the spring meeting of the Materials Research Society, San Francisco, California, April 4.

Berman, R., and M. Martinez. 1976. Diamond Research. Industrial Diamond Review.

Berzelius, J.J. 1824. Annalen der Physik. B77:209.

Bhatnagar, M., P.K. McLarty, and B.J. Baliga. 1992. Silicon-carbide high-voltage (400 V) Schottky barrier diodes. IEEE Electron Device Letters. 13(10):501-503.

Binari, S.C., L.B. Rowland, G. Kelner, W. Kruppa, H.B. Dietrich, K. Doverspike, and D.K. Gaskill. 1994. DC, RF, and High-Temperature Characteristics of GaN FET Structures. Paper presented at the International Conference on Compound Semiconductors, San Diego, California, September 18-22.

Bonn, P.A., and D.W. Palmer. 1980. 275 °C thick film hybrid microcircuitry fabrication technology. Sandia National Laboratories Report, SAND80-0078, UC-66b, March/July.

Bose, B.K. 1993. Power electronics and motion control-technology status and recent trends. IEEE Transactions on Industry Applications 29(5):902-909.

Bottner, T., K. Fricke, A. Goldhorn, H.L. Hartnagel, A. Rappl, S. Ritter, and J. Wurfl. 1991. Technology and performance of a high temperature stable operational amplifier on GaAs. Pp. 77-84 in Proceedings of the First International High-Temperature Electronics

Conference, Albuquerque, New Mexico, June 16-20.

Brown, D.M. 1993. Ion implantation and SiC Device Research and Development. Presentation to the Committee on Materials for High-Temperature Semiconductors, Washington, D.C., September 29.

Brown, D.M., G. Gati, M. Ghezzo, J. Kretchmer, V. Krishnamurthy, and G. Michon. 1994. High Temperature Silicon Carbide Planar IC Technology and First Monolithic SiC Operational Amplifier IC. Paper presented at the Second International High-Temperature Electronics Conference, Charlotte, North Carolina, June 5-10.

Brown, R.B. 1991. Short Course Notes. First International High Temperature Electronics Conference, Albuquerque, New Mexico, June 16.

Brown, R.B., F.L. Terry, and K-C Wu. 1994. High temperature microelectronics—expanding the applications for smart sensors. Pp. 274-277 in Proceedings of the International Electron Devices Meeting. New York: IEEE Press.

Chelikowsky, J.R., and S.G. Louie, 1984. First principles combination of atomic orbitals method for the cohesive and structural properties of solids—application to diamond. Physical Review B—Condensed Matter 26(6):3470-3481.

Chiao, Y.H., A.K. Knudsen, and I.F. Hu. 1991. Interfacial bonding in brazed and cofired aluminum nitride. ISHM Proceedings Pp. 460-468.

Chien, F.R., S.R. Nutt, W.S. Yoo, T. Kimoto, and H. Matsunami. 1994. Terrace growth and polytype development in epitaxial beta-SiC films on alpha-SiC (6H and 15R) substrates. Journal of Materials Research 9(4):940-954.

Chitale, S.M., C. Huang, and S.J. Sten. 1994. ESL thick-film materials for AlN. Advancing Microelectronics 21(1):22-23.

Chow, T.P., and R. Tyagi. 1994. Wide bangap compound semiconductors for superior high-voltage unipolar power devices. IEEE Transactions on Electron Devices 41(8):1481-1483.

Choyke, W.J., and I. Linkov. 1993. A short atlas of luminescence and absorption lines and bands in SiC, GaN, AlGaN and AlN. Pp. 141-146 in Proceedings of the 5th International Conference on Silicon-Carbide and Related Materials. M.G. Spencer, R.P. Devaty, J.A. Edmond, M. Asif Khan, R. Kaplan, and M. Rahman, eds. Institute of Physics Conference Series #137. Bristol, England: Institute of Physics Publishing.

Christenson, D. 1991. High temperature electronics for supersonic aircraft. Final Report on Workshop on High Temperature Electronics, June 6-8. 1989, Albuquerque, New Mexico. SAND91-0370. Albuquerque, New Mexico: Sandia National Laboratories.

Chu, T.L., D.W. Ing, and A.J. Noreika. 1967. Epitaxial growth of aluminum nitride. Solid State Electron 10(1023):67.

Clarke, R.C., R.H. Hopkins, C.D. Brandt, M.C. Driver, D.L. Barrett, A.A. Burk, G.W. Eldridge, H.M. Hobgood, J.P. McHugh, P.G. McMullin, R.R. Siergiej, and S. Sriram. 1993. Paper presented at the 1993 IEEE/Cornell Conference on Advanced Concepts in High Speed Semiconductor Devices and Circuits.

Crofton, J.B., C.S. Patuwathavithane, P.A. Barnes, J.R. Williams, M.J. Bozack, C.C. Tin, J.A. Spitznagel, P.G. McMullin, D.L. Barrett, R.H. Hopkins, and R.G. Seidensticker. 1991. Metallization Studies at Elevated Temperatures on SiC Substrates and Epitaxial Layers. Paper presented at the First International High Temperature Electronics Conference, Albuquerque, New Mexico, June 16-20.

Crofton, J.B., J.R. Williams, M.J. Bozack, and P.A. Barnes. 1993. A TiW high-temperature Ohmic contact to n-type 6H-SiC. Pp. 719-722 in Proceedings of the Fifth Conference on Silicon Carbide and Related Materials. M.G. Spencer, R.P. Devaty, J.A. Edmond, M. Asif Khan, R. Kaplan, and M. Rahman, eds. Institute of Physics Conference Series #137. Bristol, England: Institute of Physics Publishing.

Crofton, J.B., J.R. Williams, M.J. Bozack, and E.D. Luckowski. 1994. The Effect of Silicide Stoichiometry on the Ni Ohmic Contact to n-

Type 6H-Silicon Carbide. Paper presented at the Second International High Temperature Electronics Conference, Charlotte, North Carolina, June 5-10.

Davies, G. 1994. Optical spectroscopy of defects in diamond: current understanding and future problems. Pp. 21-28 in Proceedings of the 17th International Conference on Defects in Semiconductors, Gmunden, Austria, July 18-23.

Davis, R.F. 1992. Diamond Films and Coatings. Park Ridge, N.J.: Noyes Publications.

Davis, R.F. 1993. Bulk crystals, thin fims and devices of the wide band gap semiconductors of silicon carbide and the III-V nitrides of aluminum, gallium and indium. Pp. 1-6 in Proceedings of the Fifth Conference on Silicon-Carbide and Related Materials. M.G. Spencer, R.P. Devaty, J.A. Edmond, M. Asif Khan, R. Kaplan, and M. Rahman, eds. Institute of Physics Conference Series #137. Bristol, England: Institute of Physics Publishing.

Davis, R.F., G. Kelner, M. Shur, J.W. Palmour, and J.A. Edmond. 1991. Thin film deposition and microelectronic and optoelectronic device fabrication and characterization in monocrystalline alpha and beta silicon carbide. Proceedings of the IEEE 79(5): 677-701.

Deal, B.E., and A.S. Grove. 1965. General relationship for thermal oxidation of silicon. Journal of Applied Physics 36(3770):65.

Dell'Acqua, R., and M. Marelli. 1990. Hybrid circuits in automotive applications. Hybrid Circuit Technology 7(5):22-29.

Dimitriev, V.A., K.G. Irvine, M.G. Spencer, and I.P. Nikitina. 1994. Heteroepitaxial growth of SiC on AlN by chemical vapor deposition. Pp. 67-70 in Proceedings of the Fifth Conference on Silicon Carbide and Related Materials. M.G. Spencer, R.P. Devaty, J.A. Edmond, M. Asif Khan, R. Kaplan, and M. Rahman, eds. Institute of Physics Conference Series #137. Bristol, England: Institute of Physics Publishing.

Draper, B.L., and D.W. Palmer. 1979. Extension of high-temperature electronics. IEEE Transactions on Components, Hybrids, and Manufacturing Technology CHMT-2(4):399-404.

Eden, R. 1994. Gallium arsenide and high-temperature packaging. Presentation to the Committee on Materials for High-Temperature Semiconductor Devices. Washington, D.C., February 10-11.

Edmond, J.A., D.G. Waltz, S. Brueckner, H.S. Kong, J.W. Palmour, and C.H. Carter, Jr. 1991. High temperature rectifiers in 6H-silicon carbide. Pp. 499-505 in Proceedings of the First International High Temperature Electronics Conference, D.B. King and F.V. Thome, eds.

Efremow, N.N., M.W. Geiss, D.C. Flanders, G.A. Lincoln, and N.P. Economou. 1985. Ion-beam assisted etching of diamond. Journal of Vacuum Science & Technology B3(1):416-418.

Fazi, C., M. Dudley, S. Wang, and M. Ghezzo. 1993. Issues associated with large area SiC diodes with avalanche breakdown. Pp. 487-490 in Proceedings of the 5th International Conference on Silicon-Carbide and Related Materials. M.G. Spencer, R.P. Devaty, J.A. Edmond, M. Asif Khan, R. Kaplan, and M. Rahman, eds. Institute of Physics Conference Series #137. Bristol, England: Institute of Physics Publishing.

Fendrock, J.J., and L.M. Hong. 1990. Parallel-gap welding to very-thin metallization for high temperature microelectronic interconnects. IEEE Transactions on Components, Hybrids, and Manufacturing Technology 13(2):376-382.

Foresi, J.S., and T.D. Moustakas. 1993. Metal contacts to gallium nitride. Applied Physics Letters 62(22):2859-2861.

Foyt, A.G. 1994. High Temperature Electronic Materials Device Testing. Presentation to the Committee on Materials for High-Temperature Semiconductor Devices, Washington, D.C., February 10-11.

Frank, R., and R. Valentine. 1990. Power FETs cope with the automotive environment. PCIM February:33-39.

Fung, C.D., and J.J. Kopanski. 1984. Thermal oxidation of 3C silicon-carbide single crystal layers on silicon. Applied Physics Letters 45(7):757-759.

Ghezzo, M., D.M. Brown, E. Downey, J. Kretchmer, W. Hennessy, D.L. Polla, and H. Bakhru. 1992. Nitrogen-implanted SiC diodes using high-temperature implantation. IEEE Electron Device Letters 13(12)639-641.

Ghezzo, M., D.M. Brown, E. Downey, J. Kretchmer, and J.J. Kopanski. 1993. Boron-implanted 6H-SiC diodes. Applied Physics Letters 63(9):1206-1208.

Grzybowski, R.R. 1991. Development of 600 °C Device Test Fixturing. Transactions of the First International High Temperature Electronics Conference, Albuquerque, New Mexico, June 16-20.

Grzybowski, R.R., and S.M. Tyson. 1993. High temperature testing of SOI devices to 400°C. Pp. 176-177 in IEEE International SOI Conference Proceedings, Palm Springs, California, October 5-7.

Harman, G.G. In Press. Introduction to diffusion, intermetallic compounds and other metallurgical problems in high-temperature electronics with emphasis on first-level packaging. In High-Temperature Electronics. R.K. Kirscham, ed. New York: IEEE Press.

Hemstreet, L.A., and C.Y. Fong. 1974. Pp. 284-297 in Silicon Carbide. 1973. R.C. Marcshall, J.W. Faust, and C.E. Ryan, eds. South Carolina: University of South Carolina Press.

Hingorani, N.G., and K.E. Stahlkopf. 1993. High power electronics. Scientific American 269(5):78-85.

Hobgood, H.M. 1993. Growth of large diameter SiC crystals. Presentation to the Committee on the Study of Materials for High-Temperature Semiconductor Devices, Washington, D.C., September 29-30.

Ivanov, P.A., and V.E. Chelnokov. 1992. Recent developments in SIC single-crystal electronics. Semiconductor Science and Technology 7:863-880.

Johnson, E.O. 1965. Physical limitations on frequency and power parameters of transistors. RCA Review 26:163-177.

Johnson, R.W. In Press. Hybrid assembly and packaging. In High Temperature Electronics. R.K. Kirscham, ed. New York: IEEE Press.

Jurgens, R.F. 1982. High-temperature electronics applications in space exploration. IEEE Transactions on Industrial Electronics 29(2):107-111.

Kang, S., P. Neudeck, J. Petit, and M. Tabib-Azar. 1993. Measurement of fast and slow interface traps in n-type dry thermally oxidized 6H-SiC MOS diodes by high-frequency and quasi-static C-V techniques. Pp. 633-636 in Proceedings of the 5th International Conference on Silicon-Carbide and Related Materials. Philadelphia: Institute of Physics Publishing.

Keith, E. 1990. Future packaging technologies for hybrid electronics in the automotive underhood environment. Hybrid Circuit Technology 7(5):12-16.

Kelner, G., M.S. Shur, S. Binari, K.J. Sleger, and H.S. Kong. 1989. High-transconductance beta SiC buried gate JFETS. IEEE Electron Devices Letters 36(6):1045-1049.

Keyes, R.W. 1972. Figure of Merit for semiconductors for high-speed switches. Proceedings of the IEEE 60(2):225.

Khan, M.A., A. Bhattarai, J.N. Kuznia, and D.T. Olson. 1993a. High-electron-mobility transistor based on a GaN-AlXGAl-XN heterojunction. Applied Physics Letters 63(9):1214-1215.

Khan, M.A., J.N. Kuznia, A.R. Bhattra, and D.T. Olson. 1993b. Metal-semiconductor field effect transistor based on single crystal GaN. Applied Physics Letters 62(15):1786-1787.

Khan, M.A., J.N. Kuznia, D.T. Olson, W.J. Schiff, and J. Burn. In Press. Microwave.

Knippenberg, W.F. 1963. Philips Research Reports 18(3):161-274.

Knoll, G.F. 1989. Radiation Detection and Measurements, 2nd ed. New York: John Wiley & Sons.

Koga, K., Y. Fujikawa, Y. Ueda, and T. Yamaguchi. 1992. Growth and characterization of 6H-SiC bulk crystals by the sublimation method. Pp. 96–100 in Amorphous and Crystalline Silicon Carbide IV, Vol. 71. C.Y. Yang, M.M. Rahman, and G.L. Harris, eds. Berlin and Heidelberg: Springer-Verlag.

Kordina, O., J.P. Bergman, A. Henry, and E. Janzén. 1995. Long minority carrier lifetimes in 6H-SiC grown by chemical vapor deposition. Applied Physics Letters 66(22):189-191.

Kueser, P.E. 1965-1966. Development and evaluation of magnetic and electrical materials capable of operating in the 800-1600 F temperature range.

References

Westinghouse Electric Corporation Quarterly Reports. NASA-CR-54354 to 60.

Lambrecht, W.R., and B. Segall. 1992. A comparison of wurtzite and zincblende band structures for SiC, AlN, and GaN. Pp. 367-373 in the Proceedings of the Materials Research Society Symposium on Wide Bandgap Semiconductors. T.D. Mostaksas, J.I. Pankove, and Y. Hamakawa, eds. Philadelphia: Elsevier Science.

Larkin, D.J., P.G. Neudeck, J.A. Powell, and L.G. Matus. 1993. Site-competition epitaxy for controlled doping of CVD silicon carbide. Pp. 155-159 in Proceedings of the 5th International Conference on Silicon-Carbide and Related Materials. M.G. Spencer, R.P. Devaty, J.A. Edmond, M. Asif Khan, R. Kaplan, and M. Rahman, eds. Institute of Physics Conference Series #137. Bristol, England: Institute of Physics Publishing.

Laukhe, Y., Y.M. Tairov, V.F. Tsvetkov, and F. Schepanksi. 1981. Oxidation-kinetics of SiC single-crystals. Inorganic Materials 17(2):177-179.

Lee, R., C. Ito, B. Johnson, G. Trombley, R. Reston, M. Mah, and C. Havasy. 1995. High-temperature characteristics of GaAs MESFET devices fabricated with AlAs buffer layer. IEEE Electron Device Letters 16(6).

Lely, J.A. 1955. Darstellung von einkristallen von silicium carbid und beherrschung von art und menge der eingebautem verunreinigungen. Ber. Deut. Keram. Ges. 32:229-236.

Licari, J.J., and L.R. Enlow. 1988. Hybrid Microcircuit Technology Handbook. Park Ridge, New Jersey: Noyes Publications.

Lin, M.E., Z.F. Fan, L.H. Allen, and H. Morkoc. 1994. Low resistance contacts on wide-bandgap GaN. Applied Physics Letters 64(8):1003-1005.

Lo, T.C., and H-C. Huang. 1992. Japanese Journal of Applied Physics 31(1):4061.

Look, D. 1989. Electrical Characterization of GaAs Materials and Devices. New York: John Wiley & Sons.

MacKay, C.A. 1991. Bonding amalgam and method making. United States patent 5,053,195. October 1.

Maier, K, H.D. Müller, and J. Schneider. 1992. Materials Science Forum 83-87:1183-1194.

Marsh, O.J., and H.L Dunlap. 1970. Radiation Effects 6:301.

Matsunami, H., K. Shibahara, N. Kuroda, W. Yoo, and S. Nishino 1989. VPE growth of SiC on step-controlled substrates. Pp. 34-59 in Amorphous and Crystalline Silicon Carbide. G.L. Harris and C.Y.-W. Yang, eds. Berlin and Heidelberg: Springer-Verlag.

McGarrity, J.M., R.E. Oakley, W.M. Delancey, and F.B. McLean. 1992. Displacement damage effects in SiC JFETs as a function of temperature. Proceedings of the 10th Symposium on Space and Nuclear Power and Propulsion.

Migitaka, M., and K. Kurachi. 1994. Silicon integrated injection logic operating up to 454 °C. Pp. 26-32 in Transactions of the 2nd International High Temperature Electronics Conference. D.B. King and F.V. Thome, eds. Charlotte.

Miller, T. 1987. Small motor drives expand their technology horizons. Power Engineering Journal September:283-289.

Moisson, H. 1904. Compt Rend. 139:773-780.

Moisson, H. 1905. Compt Rend. 140:405-406.

Morkoc, H., S. Strite, G.B. Gao, M.E. Lin, B. Sverdlov, and M. Burns. 1994. A review of the large bandgap SiC, III-V nitride, and ZnSe based II-VI semiconductor device technologies. Journal of Applied Physics Review 76(3):1363-1398.

Motz, P.R., and W.A. Vincent. 1984. Automotive electronics: designing custom ICs for a harsh environment. Journal of Semicustom ICs 2(2):5-12.

Nakamura, S. 1991. GaN growth using GaN buffer layer. Japanese Journal of Applied Physics 30:1998.

Nakamura, S., N. Iwasa, M. Seno, and T. Mukai. 1992. Hole compensation mechanism of P-type GaN films. Japanese Journal of Applied Physics 31(5A):1258-1266.

Neudeck, P.G., and J.A. Powell. 1994. Performance limiting micropipe defects in silicon carbide wafers. IEEE Electron Device Letters 15(2):63-65.

Neudeck, P.G., D.J. Larkin, J.E. Starr, J.A. Powell, C.S. Salupo, and L.G. Matus. 1993. Greatly improved 3C-SiC p-n junction diodes grown by chemical vapor deposition. IEEE Electron Device Letters 14(3):136-139.

Nieberding, W.C., and J.A. Powell. 1982. High-temperature electronic requirements in aeropropulsion systems. IEEE Transactions on Industrial Electronics IE-29(2):103-106.

Nishino, S., J.A. Powell, and H.A. Will. 1983. Production of large-area single-crystal wafers of cubic SiC for semiconductor devices. Applied Physics Letters 42(5):460-462.

NRC (Nuclear Regulatory Commission). 1993. Research Publication GR-0005, Volume 2, Part 1. Washington, D.C.: U.S. Department of Energy.

Padiyath, R., R.L. Wright, M.I. Chaudhry, and S.V. Babu. 1991. Reactive ion etching of monocrystalline, polycrystalline, and amorphous-silicon carbide in CF_4/O_2 mixtures. Applied Physics Letters 58(10):1053-1055.

Palmer, D.W. In Press. High-temperature electronics packaging. In High Tempemperature Electronics. R.K. Kirscham, ed. New York: IEEE Press.

Palmer, D.W., and R.C. Heckman. 1978. Extreme temperature range electronics. IEEE Transactions of Components, Hybrids, and Manufacturing Technology CHMT-1:333-340.

Palmour, J.W. 1993. Design and Fabrication of SiC Devices. Presented to the Committee on Materials for High Temperature Semiconductor Devices, Washington, D.C., September 30.

Palmour, J.W., S. Kong, D.G. Waltz, J.A. Edmond, and C.H. Carter. 1991. 6H-SiC transistors for high temperature operation. Pp. 511-518 in Transactions of the First International High Temperature Electronics Conference.

Pan, W.Z., and A.J. Steckl. 1990. Journal of the Electrochemical Society 137:212.

Pandy, D., and P. Krishna. 1983. Pp. 213-258 in Crystal Growth and Characterization of Polytype Structures, P. Krishna, ed. London: Pergamon Press.

Parsons, J.D. 1987. Single crystal epitaxial growth of beta-SiC for device and integrated circuit applications. Pp. 271-282 in Novel Refractory Semiconductors. D. Emin, T.L. Aselage, and C. Wood, eds. Pittsburgh, Pennsylvania: Materials Research Society.

Patrick, L.A., and W.J. Choyke. 1974. Physics Review B 10:5091.

Pearton, S.J., C.R. Abernathy, F. Ren, J.R. Lothian, P.W. Wisk, and A. Katz. 1993. Dry and wet etching characteristics of InN, AlN, and GaN deposited by electron-cyclotron resonance metalorganic molecular beam epitaxy. Journal of Vacuum Science & Technology 11(4):1772-1775.

Pedder, D.J. 1988. Flip chip solder bonding for microelectronic application. Hybrid Circuits 15:4-11.

Pensl, G., and W.J.Choyke. 1993. Electrical and optical characterization of SiC. Physica B 185(1-4):264-283.

Perlin, A. 1993. Physical properties of single crystals of III-V nitrides grown by a high-pressure, high-temperature method. Bulletin of the American Physical Society 38(1):445.

Petit, J.B., P.G. Neudeck, L.G. Matus, and J.A. Powell. 1992. Thermal oxidation of single-crystal silicon carbide: kinetic, electrical and chemical studies. Pp. 190-196 in Proceedings of Amorphous and Crystalline Silicon Carbide IV, Springer Proceedings in Physics, Vol. 71. New York: Springer-Verlag

Plano, M.A., M.D. Moyer, and M.M. Moreno. 1994. CVD diamond MESFET. Paper presented at the Second International High Temperature Electronics Conference, Charlotte, North Carolina, June 5-10.

Porter, L.M., R.F. Davis, J.S. Bow, M.J. Kim, and R.W. Carpenter. 1993. Deposition and characterization of Schottky and ohmic contacts on n-type alpha (6H)-SiC (0001). Pp. 581-584 in Proceedings of the 5th International Conference on Silicon-Carbide and Related Materials. M.G. Spencer, R.P. Devaty, J.A. Edmond, M. Asif Khan, R. Kaplan, and M. Rahman, eds. Institute of Physics Conference Series #137. Bristol, England: Institute of Physics Publishing.

Porter, L.M., R.F. Davis, J.S. Bow, and M.J. Kim. 1995a. Chemistry, microstructure, and electrical properties at interfaces between thin films of cobalt and alpha (6H) silicon carbide (0001). Journal of Materials Research 10(1):26-33.

Porter, L.M., R.F. Davis, J.S. Bow, and M.J. Kim. 1995b. Chemistry, microstructure, and electrical properties at interfaces between thin films of titanium and alpha (6H) silicon carbide (0001). Journal of Materials Research 10(3):668-679.

Powell, J.A. 1993. Presentation to the Committee on Materials for High-Temperature Semiconductors, Washington, D.C.

Powell, J.A., and L.G. Matus. 1989. Pp. 2-12 in Springer Proceedings in Physics. G.L. Harris and C.Y-W Yang, eds. Berlin and Heidelberg: Springer-Verlag.

Powell, J.A., L.G. Matus, and M.A. Kuczmarski. 1987. Growth and characterization of cubic SiC single-crystal films, Si. Journal of the Electrochemical Society 134(6):1558-1565.

Powell, J.A., J.B. Petit, J.H. Edgar, I.G. Jenkins, L.G. Matus, J.W. Yang, P. Pirouz, W.J. Choyke, L. Clemen, and M. Yoganathan. 1991. Controlled growth of 3C-SiC and 6H-SiC films on low-tilt-angle vicinal (0001) 6H-SiC wafers. Applied Physics Letters 59(3):333-335.

Powell, J.A., D.J. Larkin, P.G. Neudeck, J.W. Yang, and P. Pirouz. 1994. Investigation of defects in epitaxial 3C-SiC, 4H-SiC and 6H-SIC films grown on SiC substrates. Pp. 161-164 in Silicon Carbide and Related Materials: Proceedings of the Fifth International Conference. M.G. Spencer, R.P. Devaty, J.A. Edmond, M.A. Kahn, R. Kaplan, and M. Rahman, eds. Bristol, England: Institute of Physics Publishing.

Prinz, J. 1994. Presentation at the Spring Meeting of the American Physical Society, Pittsburgh, Pennsylvania.

Ridley, B.K. 1993. Quantum Processes in Semiconductors. Oxford, England: Oxford University Press.

Rivard, J. 1986. Schematic of a hypothetical drive-by-wire system for an automobile with computerized traction control steering and suspension. Pp. 20-22 in Proceedings of the International Congress on Transportation Electronics. Warrendale, Pennsylvania: Society of Automotive Engineers.

Roesch, W.J. 1988. Gallium arsenide IC reliability. Tutorial Presentation at the International Reliability Physics Symposium. Monterey, California. April.

Rutz, R.F. 1976. Ultraviolet electroluminescence in AlN. Applied Physics Letters 28(7):379-381.

Sampson, R.N., and D.N. Mattox. 1991. Materials for electronic packaging. In Electronic Packaging and Interconnection Handbook. Ch. A. Harper, ed. New York: McGraw-Hill, Inc.

Schadt, M., G. Pensl, R.P. Devaty, W.J. Choyke. 1994. Anisotropy of the electron Hall mobility in 4H, 6H, and 15R silicon carbide. Applied Physics Letters 65(24):3120-3122.

Schaffer, W.J., H.S. Kong, G.H. Negley, and J.W. Palmour. 1994. Pp. 155-159 in Proceedings of the 5th International Conference on Silicon-Carbide and Related Materials. M.G. Spencer, R.P. Devaty, J.A. Edmond, M. Asif Khan, R. Kaplan, and M. Rahman, eds. Institute of Physics Conference Series #137. Bristol, England: Institute of Physics Publishing.

Shaikh, A. 1994. Thick-film pastes for AlN substrates. Advancing Microelectronics 21(1):18-21.

Shenai, K., R.S. Scott, and B.J. Baliga. 1989. Optimum semiconductors for high-power electronics. IEEE Transactions on Electron Devices 36(2):1811-1823.

Shenoy, K.V., C.G., Fonstad, Jr., and J.M. Mikkelson. 1994. High temperature stability of refractory-metal VLSI GaAs MESFETs. IEEE Electron Device Letters 15(3):106-108.

Shor, J.S., L. Bemis, and A.D. Kurtz. 1994. Characterization of monolithic n-type 6H-SiC piezoresistive sensing elements. IEEE Transactions on Electron Devices 41(5):661-665.

Shur, M., B. Gelment, C. Saavedra-Munoz, and G. Kelner. 1993. Potential of wide bandgap devices for high-temperature applications. Pp. 465-470 in Proceedings of the 5th International Conference on Silicon-Carbide and Related Materials, Washington, D.C., November 1-3. Philadelphia, Pennsylvania: Institute of Physics Publishing.

Sinclair, P. 1979. Service company needs. Pp. 27-38 in High Temperature Electronics and Instrumentation Seminar Proceedings, Houston, Texas, December 3-4.

Singh, N., and A. Rys. 1993. Thermal oxidation and electrical properties of silicon carbide metal-oxide-semiconductor structures. Journal of Applied Physics 73(3):1279-1283.

Skira, C.A., and M. Agnello. 1992. Control systems for the next century's fighter engines. Journal of

Engineering for Gas Turbines and Power 114(October):749-754.

Slack, G.A. 1964. Thermal conductivity of pure and impure silicon, silicon carbide, and diamond. Journal of Applied Physics 35(12):3460.

Slack, G.A., and S.F. Bartram. 1975. Thermal expansion of some diamond-like crystals. Journal of Applied Physics 46(1)89-98.

Spear, K.E., and J.P. Dismukes. 1994. Synthetic Diamond. New York: John Wiley & Sons.

Spitznagel, J.A. 1994. Personal communication to W.J. Choyke. Rough Cost Estimates for SiC Electronics in Nuclear Applications. May 28.

Strite, S., and H. Morkoc. 1992. GaN, AlN and InN: a review. Journal of Vacuum Science and Technology B 10(4):1237-1266.

Sze, S.M. 1981. Physics of Semiconductor Devices, 2nd ed. New York: John Wiley & Sons.

Tairov, Y.M., and V.E. Chelnekov. 1994. Personal Communication to W.J. Choyke.

Tairov, Y.M., and V.F. Tsvetkov. 1978. Journal of Crystal Growth 43:209.

Tairov, Y.M., and V.F. Tsvetkov. 1981. Journal of Crystal Growth 52:146.

Tairov, Y.M., and V.F. Tsvetkov. 1983. Progress in controlling the growth of polytypic crystals. Pp. 111-162 in Crystal Growth and Characterization of Polytype Structures. P. Krishna, ed. Oxford, England: Pergamon Press.

Tairov, Y.M., I.I. Khlebnikov, and V.F. Tsvetkov. 1974. Investigation of silicon-carbide single crystals doped with scandium. Phys Stat Sol. 25(1):349-357.

Takahashi, J., M. Kanaya, and Y. Fujiwara. 1994. Sublimation growth and characterization of SiC single crystalline ingots on faces perpendicular to a (0001) basal plane. Pp. 13-16 in Silicon Carbide and Related Materials: Proceedings of the Fifth International Conference. M.G. Spencer, R.P. Devaty, J.A. Edmond, M.A. Kahn, R. Kaplan, and M. Rahman, eds. Bristol, England: Institute of Physics Publishing.

Thornton, R.D. 1992. Power electronics for propulsion. Pp. 3-10 in Proceedings of the 7th Annual Applied Power Electronics Conference, APEC'92, Boston, Massachusetts.

Tillman, K.D., and T.J. Ikeler. 1992. Integrated flight/propulsion control for flight critical applications: a propulsion system perspective. Journal of Engineering for Gas Turbines and Power 114(October):755-762.

Tomana, M., R. Johnson, R. Wayne, R.C Jaeger, and W.C. Dillard. 1993. A hybrid silicon carbide differential amplifier for 350°C operation. IEEE Transactions on Components, Hybrids, and Manufacturing Technology 16(5):536-542.

Urushidani, T., S. Kobayashi, T. Kimoto, and H. Matsunami. 1993. High-voltage Au/6H-SiC Schottky barrier diodes. Pp. 471-474 in Proceedings of the 5th International Conference on Silicon-Carbide and Related Materials. M.G. Spencer, R.P. Devaty, J.A. Edmond, M. Asif Khan, R. Kaplan, and M. Rahman, eds. Institute of Physics Conference Series #137. Bristol, England: Institute of Physics Publishing.

Veneruso, A.F. 1979. High temperature technology--potential, promsie, and payoff. Pp. 17-26 in High Temperature Electronics and Instrumentation Seminar Proceedings, Houston, Texas, December 3-4.

Verma, A.P., and P. Krishna. 1966. Polymorphism and Polytypism in Crystals. New York: John Wiley & Sons.

Waldrop, J.R., and R.W. Grant. 1993. Schottky-barrier height and interface chemistry of annealed metal contacts to alpha-6H-SiC-crystal-face dependence. Applied Physics Letters 62(21):2685-2687.

Waldrop, J.R., R.W. Grant, Y.C. Wang, and R.F. Davis. 1992. Metal Schottky-barrier contacts to alpha-6H-SiC. Journal of Applied Physics 72(10):4757-4760.

Wang, C., K.S. Ailey, K.L. More, and R.F. Davis. 1993. Deposition of highly resistive, undoped and p-type, magnesium-doped gallium nitride films by modified gas source molecular beam epitaxy. Pp. 417-420 in Proceedings of the 5th International Conference on Silicon-Carbide and Related Materials. M.G. Spencer, R.P. Devaty, J.A. Edmond, M. Asif Khan, R. Kaplan, and M. Rahman, eds. Institute of Physics Conference Series #137. Bristol, England: Institute of Physics Publishing.

References

Wurfl, J., B. Jankl, K.H. Rooch, and S. Thierbach. 1994. GaAs Microwave Devices Operating at High Ambient Temperatures: Technology and Performance. Paper presented at Second International High Temperature Electronics Conference, Charlotte, North Carolina, June 8.

Xie, W., J.A. Cooper, Jr., and M.R. Melloch. 1994. Monolithic NMOS digital integrated circuits in 6H-SiC. IEEE Electron Device Letters. 15(11):455-457.

Yang, J.W. 1993. SiC: problems in crystal growth and polytypic transformation. Ph.D. Thesis, Case Western Reserve University.

Appendix A: Silicon as a High-Temperature Material

Silicon is the dominant semiconductor material in use by the electronics industry today, but is generally not thought of as a *high-temperature* semiconductor material. Its comparatively narrow energy bandgap creates the majority of problems during high-temperature operation when attempting to use silicon material as a discrete device or in integrated circuits for digital, analog, or power applications. However, surveys of the literature indicate that silicon bipolar and complementary metal-oxide semiconductor (CMOS) analog and digital products can function adequately beyond the MIL SPEC limit of 125 °C. Circuit and layout techniques can extend the reliable temperature range of conventional bulk CMOS and bipolar to at least 200 °C, while a combination of bi-CMOS, conservative layout rules, supply voltage reduction, and scaling of transistor (channel) dimensions can extend the range to 250 °C. The further addition of oxide-isolated processes can extend silicon bipolar and CMOS circuitry to 300 °C by reducing leakage currents, parasitic capacitances, and threshold voltage-dependence on temperature. General high-temperature issues for semiconductors, which also pertain to silicon, are discussed in Chapter 3. This appendix begins with a description of high-temperature performance of several silicon technologies, then moves on to a consideration of oxide-isolation processes that can extend the functional temperature range of silicon circuits.

HIGH-TEMPERATURE OPERATION OF SILICON CIRCUITS

Bipolar Analog Circuits

Historically, operational amplifiers have been the most-studied bipolar analog integrated circuit. The changes in bipolar component characteristics mentioned above can be so great with respect to temperature that conventional design methods cannot be used; in fact, design compensation techniques may be only valid for limited temperature ranges. Leakage currents must be compensated for in all designs; for example, the base-collector leakage current, I_{cbo}, flows in the opposite direction to the normal base current and can become larger than the normal base current as operating temperatures increase, reducing the base current necessary to sustain collector current. Decreases in base-emitter voltage, V_{be} (less than 100 mV), can force devices to go into saturation; current design must be used to compensate for this unintentional saturation. In general, large changes in parameters such as V_{be} and diffused resistor values of the base and collector cause problems in obtaining controlled and constant circuit performance over wide ranges (Beasom and Patterson, 1982). These parameter changes manifest themselves as failures due to degradation in the input-offset voltage, V_{os}, the open-loop gain, and the bias current.

Bipolar Digital Circuits

Commercial four-input standard and Schottky-clamped TTL NAND gates were tested from 25-325 °C. The high-temperature failure modes of both TTL NAND gates were identical. The functional failure mode was low output-high voltage, V_{oh}, and contributed to the collector-base leakage current (from the phase splitter transistor) flowing through the phase splitter collector resistor. The voltage drop across the collector due to the excess leakage resulted in a decrease in V_{oh}. The power-supply currents for output-high and output remained stable through 300 °C. Current-sinking capability increased as the temperature was increased due to the increasing gain of the current-sink transistor. Current-sourcing capability was reduced due to the increase in circuit resistance values (Prince et al., 1980).

FET Analog Circuits

Design techniques for high-temperature analog CMOS generally try to address the temperature dependence of mobility, drain current, threshold voltage, and leakage currents of MOS transistors as a first attempt to increase the range of temperature operation. Observed effects in analog CMOS circuits due to changes in transistor parameters include: (1) a decrease in amplifier-gain bandwidth product and gain, (2) an increase in amplifier input-offset voltages, (3) bias point shifts, and (4) a decrease in sampling rate due to leaky switches.

For a simple two-stage CMOS operational amplifier, design techniques have been used to allow the op amp to function up to 250 °C. The most important design technique used was to bias the two gain stages and the output stage at their zero temperature coefficient (ZTC) drain currents in the saturation region (e.g., ZTC is a gate-bias voltage at which the drain current exhibits minimum temperature sensitivity). The ZTC gate bias was applied to the current source biasing each gain stage and to the n-MOS of the output stage by using a voltage dividing string composed of n- and p-MOSFETs. Leakage currents were compensated for by cascading n- and p-MOSFETs or by using compensation diodes. Cascading was used in the voltage dividing string and the output stage. Cascading of n- and p-MOSFETs to a common circuit node allows the leakage currents from the drain to body of the n-MOS and the source to body of the p-MOS to cancel each other with appropriate selection of MOS-feature sizes. For the differential input stage, a compensation diode can be placed at any node where the use of cascaded MOS pairs is not possible; the diode can then be used to shunt excess current to one of the power rails (Shoucair, 1986).

Digital CMOS

General high-temperature effects observed for digital CMOS include: (1) decreasing noise margins and (2) decreasing switching speeds. CMOS type 4012 NAND circuits were tested from 25-300 °C, with acceptable performance noted to 270 °C. Leakage current on the p-well-substrate junction was determined to limit circuit functionality. The output-low voltage, V_{ol}, increased at high temperatures due to the inability of the n-channel transistors to sink the leakage currents generated at high temperature. Change in threshold voltages also altered the logic threshold of the voltage transfer characteristics (Prince et al., 1980). Additional layout (e.g., guardrings or larger n-MOS/p-MOS separation) or processing techniques (e.g., epitaxial substrates or trenches) may be required to suppress latchup at higher temperatures (Estreich and Dutton, 1982).

DIELECTRIC ISOLATION TECHNOLOGY

Leakage current in reverse-biased junctions is one of the major problems with the operation of junction-based devices and junction-isolated integrated circuits (ICs) at high temperatures. In bulk CMOS, large junction areas exist between the source, the drain, and the p-well, and between the p-well and the substrate. For junction-isolated bipolar, large junction areas exist between the collector well and the substrate. As described above, junction-isolated circuits with no special design precautions for high temperatures can fail at temperatures as low as 200 °C.

The use of dielectric isolation (DI) or silicon-on-insulator (SOI) eliminates the problem of junction isolation in Ics by isolating each device with an oxide layer that will (1) eliminate parasitic leakage currents between devices and between devices and power rails, and (2) eliminate extra junctions that form parasitic devices. The primary benefit of SOI use for CMOS and some bipolar ICs is the elimination of latchup. SOI eliminates the formation of parasitic p-MOS devices in bipolar ICs and the formation of parasitic bipolars in CMOS ICs. Figure A-1 illustrates the reduction in large junction-isolation areas by the use of trenches and SOI.

SOI also allows MOS devices to be fabricated in a manner that reduces leakage currents within the MOS device itself. The silicon film can be made thin enough to allow the drain and source wells to contact the dielectric layer; the junction area between the source or drain and the channel is the width of the FET multiplied by the thickness of the silicon film. Reducing the thickness of the silicon film can lead to an SOI FET with leakage currents that are orders of magnitude smaller than in a bulk FET. Figure A-2 illustrates the leakage current as a function of

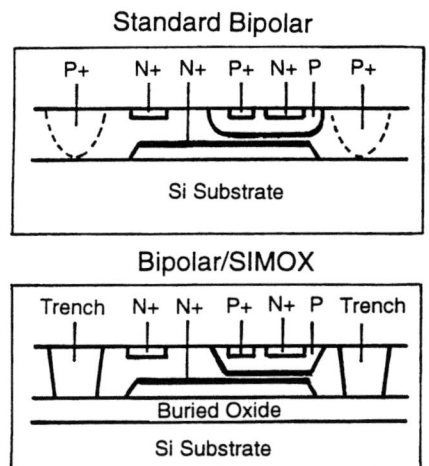

FIGURE A-1 Reduction in large junction isolation areas by the use of trenches and SOI. SOURCE: Ibis Technology Corp. (1991).

temperature for three types of n-MOS transistors with gate lengths of 2 microns. Leakage currents in thin-film transistors may not be linearly dependent on the silicon film thickness; theoretical calculations show that the leakage current decreases more rapidly as the thickness is decreased. Parasitic capacitances are also reduced by reduction of those junction areas (Swonger et al., 1991).

While the use of SOI in bipolar devices does reduce isolation, latchup, and parasitic MOS-device problems due to leakage currents at high temperatures, problems such as V_{be} reduction, transistor current gain, base current reversal, etc., still remain. These problems can be successfully addressed through circuit and device layout modifications.

DI techniques commercially used today include: (1) separation by implantation of oxygen (SIMOX); (2) wafer bonding, lapping, and etch back; and (3) V-groove etching, polysilicon filling, and lapping of the crystalline silicon.

Wafer Bonding

Wafer bonding is the latest SOI technology. Bonded wafer substrates can be prepared by thermally oxidizing two wafers. The wafers are then treated so that the oxide surfaces become hydrophilic. The oxide surfaces are then placed face-to-face, forming a weak room-temperature bond. Subsequent annealing at temperatures greater than 800 °C form stronger bonds so that the wafers can no longer be separated. After bonding, one of the wafers (the device wafer) is thinned to the desired silicon film thickness by grinding, electrochemical etching, and polishing. Thin and uniform silicon layers are difficult to produce using the wafer-bonding technique (Swonger et al., 1991). Wafer bonding is currently limited to silicon film thicknesses larger than 1 μm due to thickness variations of 0.5 μm during thinning processes. Wafer-bonding does provide an excellent quality silicon film with very few dislocations. The wafer-bonding process also facilitates the fabrication of SOI wafers with very thick buried oxide layers. The high-quality silicon films and thick oxides generally make wafer bonding a good technology for high-performance bipolar applications. Dislocations can cut through the bipolar emitter and collector, allowing preferential diffusion and punch-through. Thicker oxide layers reduce substrate capacitance, allowing higher-speed bipolar performance.

SIMOX

In the SIMOX process, a buried SiO_2 layer is formed below the silicon wafer surface by implanting oxygen into the wafer at sufficient dose and energy. The thickness and quality of the silicon and SiO_2 layers depend on the oxygen dose, temperature of the wafer during implantation, and anneal temperature after the implant. Multiple implants are used to reduce the silicon defect density. As an example, a high-quality sample was made with a 400-nm-thick buried oxide and a single-crystal 250-nm-thick silicon top film. The ion dose, energy, and implantation temperature were 1.8×10^{18} cm^{-2}, 200 keV, and 620 °C, respectively. A final post-implant anneal of 1350 °C was used. The quality is adequate enough to fabricate 256 k

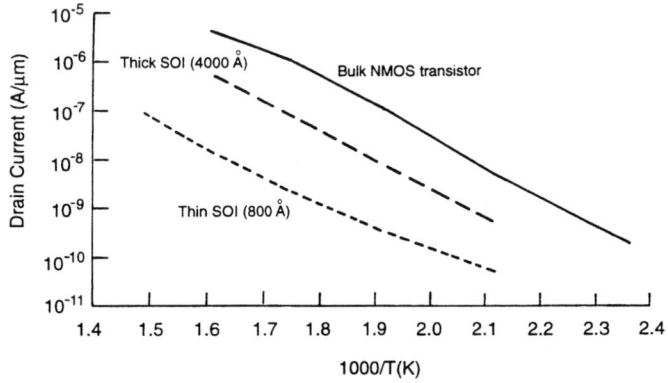

FIGURE A-2 Leakage currents as function of temperature for three types of n-MOS transistors with gate lengths of 2 microns. SOURCE: Swonger et al. (1991).

SRAMS. Commercial vendors offer SIMOX wafers from 3 in. to 200 mm diameter. SIMOX is traditionally used for CMOS applications instead of wafer bonding because thin and uniform layers can be produced (Colinge, 1993).

Lateral Isolation

In addition to vertical isolation processes afforded by the SOI technology, a lateral isolation process is needed to isolate devices. In SOI, thin silicon films can be etched off between devices or consumed by local oxidation. Lateral isolation in thick films is obtained by etching narrow and deep trenches through the silicon layer to the buried oxide layer. These trenches can then be filled with an oxide.

APPLICATIONS TO DEVICE TECHNOLOGY

Bipolar-Junction-Transistor Applications in SOI Technology

Recently, lateral n-p-n bipolar transistors have been fabricated using SIMOX SOI substrates. The bipolar structures were investigated for high-frequency and smart-power applications and no high-temperature tests were mentioned, although the benefits of dielectric isolation with respect to higher integration density, no latchup, less leakage current, high-temperature operation, and noise immunity were mentioned (Weyers et al., 1992; Parke et al., 1993).

Operational amplifiers that used dielectrically isolated bipolar transistors and other design techniques mentioned above were developed in the late 1970s by Harris (Beasom and Patterson, 1982). The operational amplifier was characterized over the temperature range of 25-300 °C. Useful circuit performance was observed up to 300 °C. In Appendix C, Table C-1 summarizes the electrical parameters of the 300 °C op amp. A similar op amp circuit design was refabricated by Harris using a DI process flow and bonded wafer SOI substrates (Swonger et al., 1991). The process flows are shown in Figure A-3. Other than dielectric isolation, no other high-temperature compensation techniques were used for this circuit design. These op amps were also tested to 300 °C, with all devices functional to that temperature. Power dissipation, input-offset voltage, input-bias current, and input-offset current increased as temperature increased while the open-loop gain decreased with increasing temperature; this latter parameter is shown in Figure A-4 as a function of temperature. It was reported that degradation could be minimized by incorporating other high-temperature design techniques into the operational amplifier design.

CMOS Applications in SOI

A variety of CMOS circuits based on SOI has been tested for high-temperature applications. Circuits tested include inverters, 19 stage ring oscillators, and 4 k through 128 k SRAMS.

Allied Signal has tested thick- and thin-film MOSFET-based inverter circuits to temperatures as high as

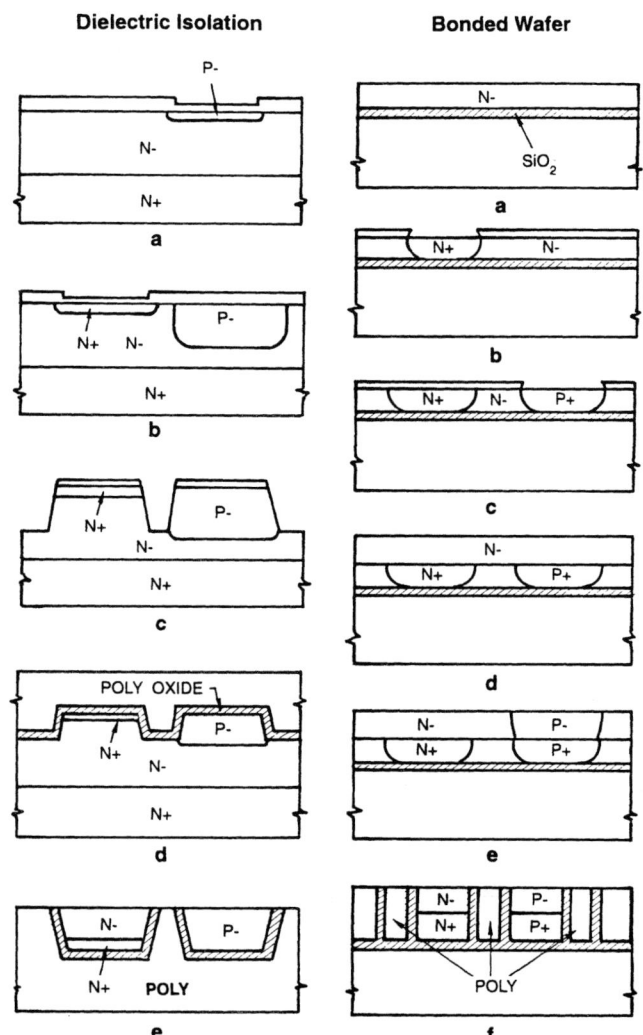

FIGURE A-3 Schematics of the dielectric isolation material process flow and the bonded wafer material process flow. SOURCE: Swonger et al. (1991).

Appendix A: Silicon as a High-Temperature Material

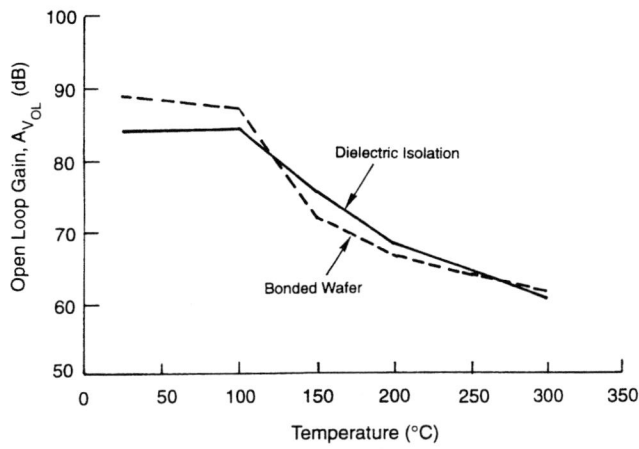

FIGURE A-4 Open-loop gain as a function of temperature. SOURCE: Swonger et al. (1991).

450 °C. The leakage current is greatly reduced by the use of SOI with two films. Inverters made in 4,000-Å-thick SOI were tested to 450 °C with fairly good results. Leakage currents did degrade their noise margin somewhat, and threshold voltage shifts did change the output-voltage versus input-voltage swing slightly (McKitterick, 1991).

Harris has tested 4 k SRAMs in silicon and SOI from 25-300 °C. The SOI SRAMs functioned to 300 °C, with degradation occurring in access times and circuit standby current. The bulk SRAMs failed at 275 °C.

SOI SRAMs (64 k) developed for military applications by Honeywell have been tested at 250 °C for 5,000 hours. The SOI CMOS process tested has not been optimized for high-temperature operation but is being modified to develop an SOI process capable of 300 °C operation. Digital products to be tested include processors and application-specific ICs as well as memories. Linear products in development for high-temperature testing include operational amplifiers, analog switches, voltage references, and application-specific integrated circuits.

REFERENCES

Beasom, J.D., and R.B. Patterson. 1982. Process characteristics and design methods for a 300 °C quadoperational amplifier. IEEE Transactions on Industrial Electronics IE-29(2):112-117.

Colinge, J.P. 1993. SOI Technology: Materials to VLSI. Amsterdam, Netherlands: Klewer Academic.

Estreich, D.B., and R.W. Dutton. 1982. Modeling latch-up in CMOS integrated circuits. IEEE Transactions on Computer-Aided Design of Integrated Circuits and Systems CAD-1(4):157-162.

Ibis Technology Corporation. 1991. SIMOX for high temperature applications. Ibis Technology Application Note, No. 103. Danvers, Massachusetts: Ibis Technology Corporation.

McKitterick, J.B. 1991. Very thin silicon-on-insulator devices for CMOS at 500 °C. Pp. 37-41 in Proceedings of the First International High Temperature Electronics Conference, Albuquerque, New Mexico, June 16-20.

Parke, S.A, C.M. Hu, and P.K. Ku. 1993. A high-performance lateral bipolar-transistor fabricated on Si MOX. IEEE Electron Device Letters 14(1):33-35.

Prince J.L., B.L. Draper, E.A. Rapp, J.N. Kronberg, and L.T. Fitch. 1980. Performance of digital-integrated-circuit technologies at very high temperatures. IEEE Transactions on Components, Hybrids, and Manufacturing Technology CHMT-3(4):571-579.

Shoucair, F.S. 1986. Design considerations in high temperature analog CMOS integrated-circuits. IEEE Transactions on Components, Hybrids, and Manufacturing Technology 9(3):242-251.

Swonger, J.W., S.J. Gaul, and P.L. Heedley. 1991. An evaluation of amp performance up to 300 °C using dielectric isolation and bonded wafer material technologies. Pp. 281-290 in Proceedings of the First International High Temperature Electronics Conference, Albuquerque, New Mexico, June 16-20.

Weyers, J., H. Vogt, M. Berger, W. Mach, B. Mutterlein, M. Raab, F. Richter, and F. Vogt. 1992. Microelectronic Engineering 19(1-4):733-736.

Appendix B: Gallium Arsenide as a High-Temperature Material

Recent advances in the quality of devices have moved this well-known semiconductor into the forefront of high-temperature electronics. In comparison to silicon, the wide bandgap of GaAs (and GaAs-based alloys) makes it an intrinsically high-temperature material (Sze, 1981). However, because the bandgap of GaAs is not as wide as that of SiC, it is not as suitable for very high temperatures (above 400 °C; Fricke et al., 1989). Over the years, GaAs FET-based high-temperature technology has developed into a well-established large-scale integrated technology, with some inroads into the very-large-scale-integrated (VLSI) arena. Although the developments in GaAs heterostructure bipolar transistors (HBTs) have also been significant, the technology for their fabrication is much less developed than that for MESFETs and heterostructure field-effect transistors (HFETS). As this technology develops, HBTs may become the preferred device for high-temperature electronics. Unlike emerging SiC technology, existing GaAs material and fabrication technology is currently able to produce integrated digital, analog, microwave, and opto-electronic circuits. The high-temperature potential of GaAs-based integrated circuit technologies is reviewed in this appendix.

STATUS OF COMMERCIAL VLSI GaAs DEVICES FOR HIGH-TEMPERATURE ELECTRONICS

The failure modes observed to date for GaAs devices have primarily been wear-out mechanisms caused by metal–GaAs interdiffusion (Christou et al., 1985; Magistrali et al., 1991). In normal operation of GaAs MESFETs with gold (most often being Ti/Pt/Au or Ti/Pd/Au) or aluminum metallization, the major modes of failure are (1) ohmic contact degradation caused by interdiffusion to the source or drain of FET structures; (2) degradation of Schottky gates caused by interdiffusion to the channel of FET structures; and (3) electromigration, usually within aluminum metallization, on surfaces (Maurer et al., 1990). When the heterostructure FETs (HFET or MODFET for modulation-doped FET), also known by many other acronyms (HEMT for high electron mobility transistor, SDHT for selectively doped heterojunction transistor, TGFET for two-dimensional electron gas FET, SISFET for semiconductor-insulator-semiconductor FET, HIGFET for heterojunction insulated-gate FET, and complementary HFETs) or HBTs are evaluated, one must add to the above MESFET failure modes interdiffusion between semiconductor layers. This can destroy the stability of the desired heterostructure (Maurer et al., 1990).

In general, MESFETs exhibit higher gate leakage currents than MOSFETs because the channel isolations from their gates are made of reverse-biased Schottky junctions in which the leakage is orders of magnitude higher than the oxides used in the latter devices. In addition, the MESFETs are likely to suffer from electron injection from the channel into the substrate because of the high electric fields generally prevailing near the drain of the devices (Shoucair and Ojala, 1992). Still, silicon transistors can only reach 200 °C if their leakage currents are properly compensated at such temperatures. In addition, the 200 °C maximum temperature for silicon devices corresponds to at least 400 °C for GaAs (Fricke et al., 1989). Recently, the performance of commercially available VLSI GaAs devices in elevated temperatures (200-400 °C) has been a subject of extensive studies (Bottner et al., 1991; Schweeger et al., 1991; Anholt and Swirhun, 1991; Simons et al., 1994). The device degradation at these elevated temperatures was attributed to drain leakage currents that caused increased output conductance, poor pinch-off characteristics, and low current-on over current-off (I_{on}/I_{off}) ratios (Lee et al., 1994). Shoucair and Ojala (1992) reported on the effects of elevated temperatures on the large- and small-signal electrical parameters of commercially available enhancement- and depletion-mode GaAs MESFETs. These MESFETs were fabricated with a tungsten nitride gate and AuGe/Ni/Au ohmic contacts. Their experimental data suggest that while GaAs MESFETs generally exhibit degradation mechanisms similar to those of silicon MOSFETs at elevated temperatures, they incur several additional effects that include (1) an increased gate leakage current; (2) a lowered Schottky barrier height; (3) a lowered sensitivity to sidegating and backgating; (4) a lowered input resistance; and (5) an increased drain resistance (Shoucair and Ojala, 1992). The authors concluded that the leakage current of commercially available GaAs MESFETS, between room temperature and 400 °C, is caused by the generation-recombination mechanism. As can be seen in Figure B-1, the leakage current varies in direct proportion to the intrinsic carrier concentration $n_i(T)$, as does the variation of GaAs substrate resistivity with temperature (Shoucair and Ojala, 1992).

Shenoy et al. (1994) recently studied the self-aligned VLSI GaAs MESFETs with tungsten-based refractory-metal Schottky gates and nickel-based refractory-metal ohmic contacts. These commercially available devices were shown to be stable after three hours at temperatures up to 500 °C (Figure B-2), with a significant degradation of the transconductance, g_m, seen above 500 °C.

Sokolich et al. (1991) described microwave FETs with an 800-hour lifetime at 250 °C, apparently limited by ohmic contact degradation. Another study by Esfandiari et al. (1990) revealed that when ion-implanted GaAs MESFETs are subject to temperatures of 125 °C for 10,000 hours, they show no degradation of the ohmic contacts and gate metallization. Fricke et al. (1989) have demonstrated that using ohmic contacts with a WSi_2 diffusion barrier and properly optimized passivated surfaces allow reliable device operation up to 300 °C, thereby proving the adaptability of state-of-the-art GaAs fabrication techniques for high-temperature applications.

APPROACHES FOR IMPROVING GaAs IC HIGH-TEMPERATURE LIMITS

In this section, different existing GaAs-based devices for high-temperature applications are described based on information provided in the literature (Hartnagel, 1992; Dreike et al., 1994). Several valuable experimental papers (Fricke et al., 1989; Anholt and Swirhun, 1991; Bottner et al., 1991; Schweeger et al., 1991; Sokolich et al., 1991; Swirhun et al., 1991; Hartnagel, 1992; Lee et al., 1994, 1995; Reston et al., 1994; Simons et al., 1994) have been published that point out the potential advantages

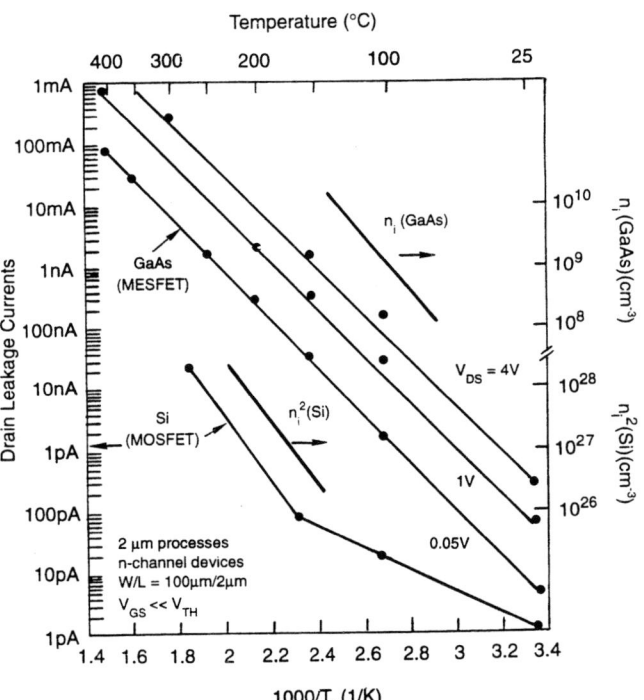

FIGURE B-1 GaAs MESFET and silicon MOSFET drain leakage currents. SOURCE: Shoucair and Ojala (1992), © 1992 IEEE.

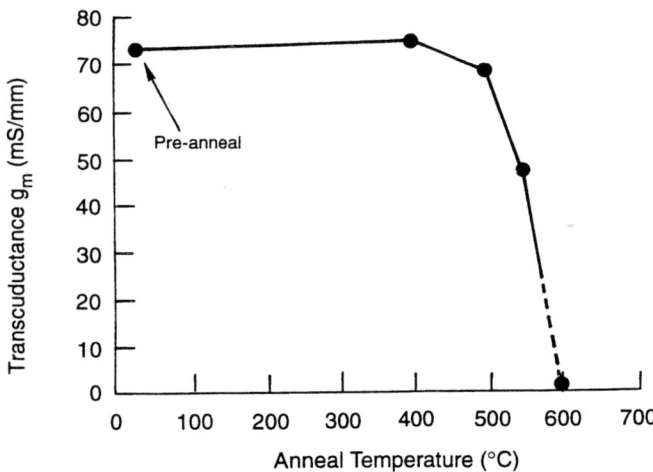

FIGURE B-2 MESFET transconductance, g_m, after three-hour anneals at various temperatures. SOURCE: Shenoy et al. (1994), © 1994 IEEE.

Appendix B: Gallium Arsenide as a High-Temperature Material

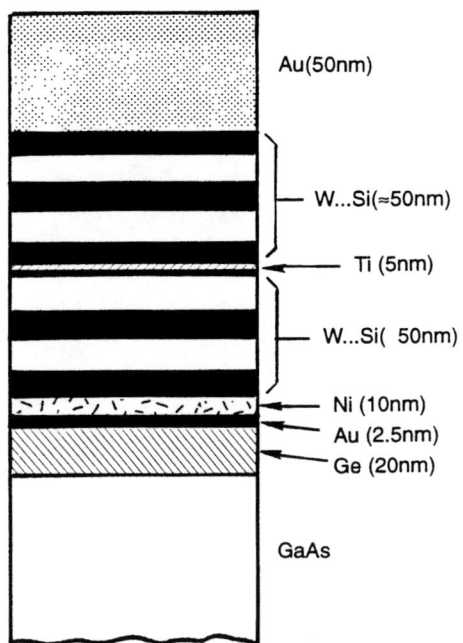

FIGURE B-3 Diffusion barrier constructed of nine alternating layers of electron-beam evaporated tungsten and silicon. SOURCE: Fricke et al. (1989), © 1989 IEEE.

of this semiconductor material. Fricke et al. (1989) reported that GaAs MESFETs experienced no device deterioration at 300 °C, even after storage, without bias at this temperature for more than 1,000 hours. The high reliability of these devices was mainly due to a diffusion barrier of WSi_2 in the ohmic contacts and an optimized Si_3N_4 passivation. The diffusion barrier was constructed of nine alternating layers of electron-beam evaporated tungsten and silicon that, after rapid thermal annealing at 640 °C, formed a 1,000-Å-thick WSi_2 layer (Fricke et al., 1989). Figure B-3 shows the structure of an ohmic contact after deposition. A simple operational amplifier constructed with these MESFETs functioned at 300 °C (Bottner et al., 1991; Schweeger et al., 1991).

Swirhun et al. (1991) demonstrated 100-hour lifetimes at 400 °C for a 1 µm self-aligned gate (SAG) MESFET with temperature-hard ohmic contacts, buried p-type channel implants, and gate sidewall spacers (Figure B-4). A SAG process-flow first defines a refractory gate metal (WSI), and then uses this gate pattern to self-align source and drain-dopant implants. This is followed by the deposition of Si_3N_4 and subsequent implants with activation at 800 °C. In this fabrication sequence, the Schottky gate and passivation layer must be mechanically and chemically stable at temperatures well above the operating range. The Ni/In/Ge/Ni/Mo ohmic contacts are made after gate definition. These ohmic contacts were passivated with 100 nm of Si_3N_4 and a rapid thermal anneal at 800 °C for five seconds.

A 1.0 µm x 10 µm depletion-mode MESFET showed on/off current ratio decreasing from 106:1 at room temperature to near 20:1 at 400 °C (Figure B-5; Swirhun et al., 1991). Although this degradation is detrimental for most electronic applications, these devices can be still used for some digital and small-signal radio frequency functions.

Lee et al. (1994, 1995) studied the influence of GaAs substrate conduction on FET drain leakage current at elevated temperatures. Other studies have shown the high resistance, undoped AlAs buffer layers to practically eliminate leakage current through the substrate. The I_{on}/I_{off} ratio for the MESFET with a 2,500-Å AlAs buffer was 330:1, which is an order of magnitude improvement over

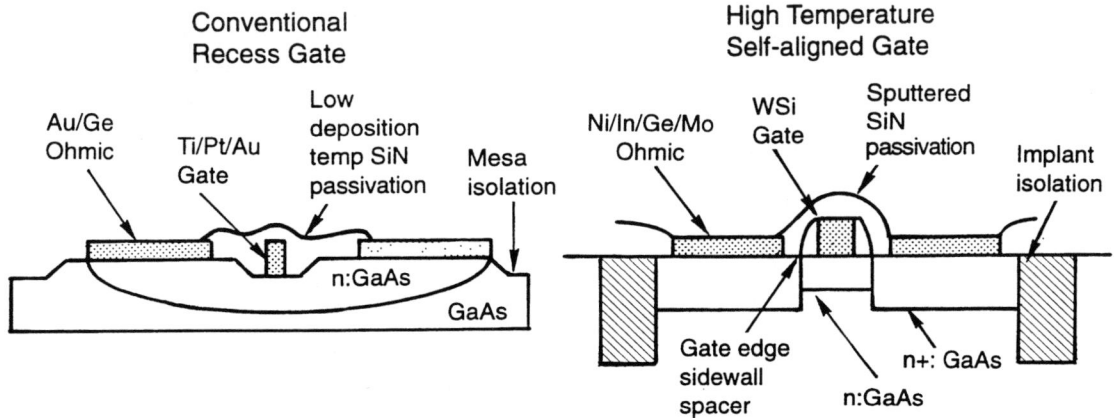

FIGURE B-4 Comparison of conventional MESFET with MESFET using temperature-hard ohmic contacts, buried p-type channel implants, and gate sidewall spacers. SOURCE: Swirhun et al. (1991).

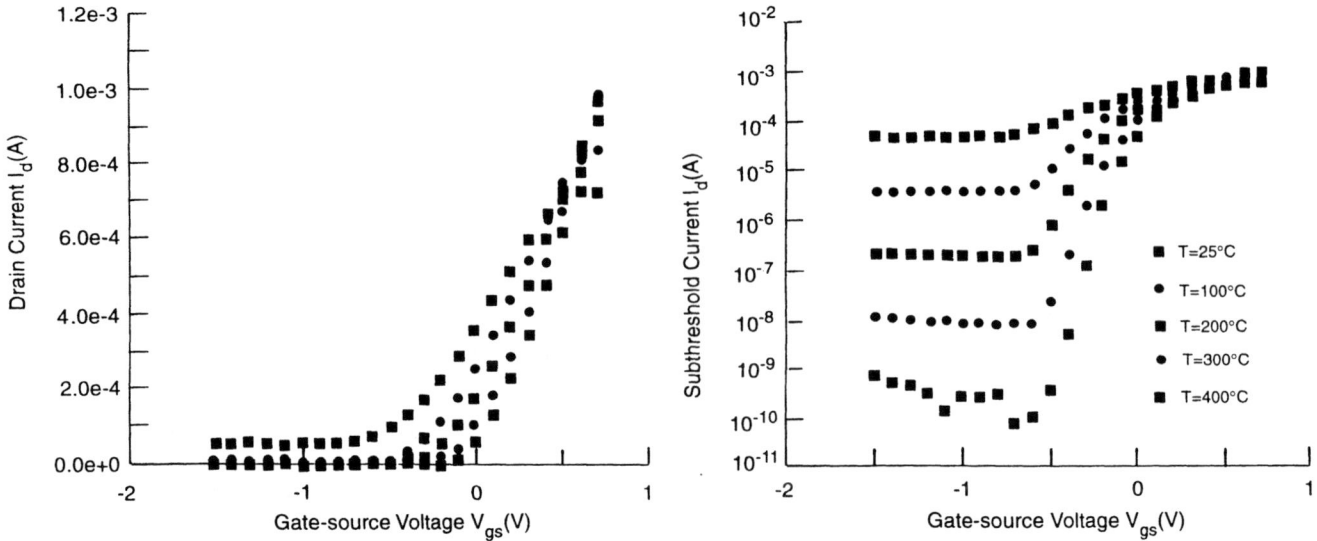

FIGURE B-5 MESFET showing on/off current ratio decreasing from 106:1 at room temperature to near 20:1 at 400 °C. SOURCE: Swirhun et al. (1991).

previously reported results at 350 °C (Simons et al., 1994; Schweeger et al., 1991). This AlAs buffer should also play a major role in reducing backgating effects for ICs. From the work of Lee and his colleagues, it seems that the reduction of I_{off} is more dependent on AlAs bulk resistance than conduction-band discontinuity. This study shows that one of the major deterrents (i.e., leakage current through the substrate) of GaAs standard technology can be removed by adding high resistivity buffer layers, thus making it a viable technology for high-temperature applications.

Reston et al. (1994) demonstrated that, through minor modifications to a standard MESFET process, the high-temperature MESFET can be fabricated (Figure B-6). The first improvement involved deposition of a silicon-nitride insulator under the interconnecting metal to reduce parasitic currents. The second modification consisted of MBE deposition of a high-resistivity AlAs buffer layer below the active device layer to reduce substrate leakage. The final change consisted of a substitution of the conventional ohmic contact with a refractory metal stack similar to the one proposed by Swirhun et al. (1991).

With these modifications, a high-temperature MESFET operating at 350 °C had its output conductance reduced by an order of magnitude (7,700 Ω at $V_g = 0V$), and the off-current reduced to approximately half of the gate leakage current. Figure B-7 shows the I-V characteristics for the typical high-temperature MESFET at 350 °C (Reston et al., 1994).

Eden (1994) advised the use of a high barrier potential gate (HiGFET) structure or possibly p-n junction gate (JFET) structure instead of a simple MESFET to reduce drain-gate leakage current and raise gate forward voltage. As an extreme to reduce the leakage, Eden suggested using the demonstrated Rockwell "lift-off" technology to transfer GaAs devices to insulating substrate, followed by isolation etch and planarization steps and then generating the metal/dielectric layers required for interconnects.

CONCLUSIONS

GaAs-based IC technologies are likely to play an important role in the realization of high-temperature devices. Both device physics and semiconductor fabrication technology demonstrate that, for selected applica-

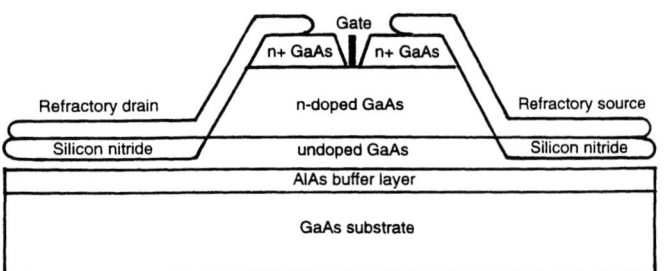

FIGURE B-6 High-temperature MESFET incorporating modifications to standard process. SOURCE: Lee et al. (1995), © 1995 IEEE.

tions, homojunction electronic GaAs devices are capable of 400 °C DC transistor characteristics and 500 °C storage without bias (Swirhun et al., 1991). With some structural and process modifications, commercially available GaAs MESFETs can be developed for utilization in high-temperature electronics (Schweeger et al., 1991; Swirhun et al., 1991; Lee et al., 1994, 1995; Reston et al., 1994). Thus, for a modest investment in process modification of commercial MESFETS, substantial high-temperature performance characteristics can be realized, and improved devices can be manufactured to support the system requirements up to 400 °C.

For GaAs-based and all other IC technologies, development of stable, electromigration-resistant metal systems for interconnecting devices in ICs and the supporting packaging technology is an important reliability issue for any high-temperature applications. To achieve this with 10^4-hour lifetimes will require further development of interconnection and package technology. With sufficient market pull, GaAs-based technology could be developed for reliable operation up to 400 °C, except for microwave devices. This technology development could be relatively straightforward and would build upon existing infrastructure. However, for the applications that demand temperatures above 400 °C, ternary and quaternary III-V material systems might offer better potential solutions (for example AlGaAs/GaAs diodes and bipolar-junction transistors grown on GaAs substrates, have demonstrated operation to 450 °C (Zipperian, 1986; Fricke et al., 1989; Dreike et al., 1994).

REFERENCES

Anholt, R., and S. Swirhun. 1991. Measurement and analysis of GaAs MESFET parasitic capacitances. IEEE Transactions on Microwave Theory Technology 39(7):1247-1251.

Bottner, T., K. Fricke, A. Goldhorn, H.L. Hartnagel, A. Rappl, S. Ritter, and J. Wurfl. 1991. Technology and performance of a high temperature stable operational amplifier on GaAs. Pp. 77-84 in Proceedings of the First International High-Temperature Electronics Conference, Albuquerque, New Mexico, June 16-20.

Christou, A., B.R. Wilkins, and W.F. Tseng. 1985. Low-temperature epitaxial growth of GaAs on (100) silicon substrates. Electronics Letters 21(9):406-408.

Dreike, P.L., D.M. Fleetwood, D.B. King, D.C. Sprauer, and T.E. Zipperian. 1994. An overview of high-temperature electronic device technologies and potential applications. IEEE Transactions on Components, Packaging, and Manufacturing Technology 17(4):594-609.

Eden, R. 1994. Gallium arsenide and high-temperature packaging. Presentation to the Committee on Materials for High-Temperature Semiconductor Devices. Washington, D.C., February 10-11.

Esfandiari, R., T.J. O'Neill, T.S. Lin, and R.K. Rono. 1990. Accelerated aging and long-term reliability study of ion-implanted GaAs MMIC if amplified. IEEE Transactions on Electron Devices 37(4):1174-1177.

Fricke, K., H.L. Hartnagel, R. Schutz, G. Schweeger, and J. Wurfl. 1989. A new GaAs technology for stable FETs at 300 °C. IEEE Electron Device Letters 10:577-579.

Hartnagel, H.L. 1992. Compound semiconductor devices for operation at elevated temperatures. Microelectronic Engineering 19:115-122.

Lee, R., C. Ito, R. Reston, G. Trombleu, B. Johnson, M. Mah, and C. Havasy. 1994. Low Leakage GaAs MESFET Devices Operating to 350 °C Ambient. Paper presented at the Second International High Temperature Electronics Conference, Charlotte, North Carolina, June 5-10.

Lee, R., C. Ito, B. Johnson, G. Trombley, R. Reston, M. Mah, and C. Havasy. 1995. High-temperature characteristics of GaAs MESFET devices fabricated

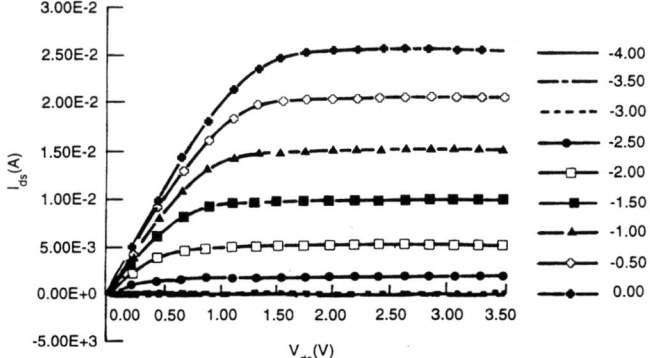

FIGURE B-7 Operating characteristics of MESFET structure shown in Figure B-6. SOURCE: Lee et al. (1995), © 1995 IEEE.

with AlAs buffer layer. IEEE Electron Device Letters 16(6).

Magistrali, F., D. Sala, M. Vanzi, F. Fantini, F. Corticelli, and A. Migliori. 1991. TEM observation of GaAs/GaAlAs laser diodes degraded in field operation. Electronics Letters 27(1):58-59.

Maurer, R.H., K. Chao, E. Nhan, R.C. Benson, and C.B. Bargeron. 1990. Reliability study of gallium arsenide transistors. Pp. 670-676 in Proceedings of the 40th Electronic Components and Technology Conference, Las Vegas, Nevada. Piscataway, New Jersey: IEEE.

Reston, R.R., H.Y. Lee, C. Ito, G.J. Trombley, Ch K. Havasy, and B. Johnson. 1994. Enhanced gallium arsenide metal-semiconductor field effect transistors designed for high temperature operation. Pp. 1138-1142 of the Proceedings of the IEEE 1994 National Aerospace and Electronics Conference, Dayton, Ohio, May 23-27. New York: IEEE Press.

Schweeger, G., K. Fricke, K. Mencke, and H.L. Hartnagel. 1991. A GaAs integrated differential amplifier for operation up to 300 °C. Solid State Electronics 34:731-733.

Shenoy, K.V., C.G. Fonstad, Jr., and J.M. Mikkelson. 1994. High temperature stability of refractory-metal VLSI GaAs MESFETs. IEEE Electron Device Letters 15(3):106-108.

Shoucair, F.S., and P.K. Ojala. 1992. High-temperature electrical characteristics of GaAs MESFETS (25-400 °C). IEEE Transactions on Electron Devices 39(7):1551.

Simons, R.N., S.R. Taub, S.A. Alterovitz, and P.G. Young. 1994. Characteristics of III-V Semiconductor Devices at High Temperature. Paper presented at the Second International High Temperature Electronics Conference, Charlotte, North Carolina, June 5-10.

Sokolich, M., K.K. Yu, M.W. Chiang, H.M. Le, and Y.C. Shih. 1991. Performance and Reliability of GaAs Refractory Gate X-band Power Amplifiers at Elevated Temperatures. Hughes Aircraft Co., Microwave Products Division. Torrance, Calif.: Hughes Aircraft.

Swirhun, S., S. Hanka, J. Nohava, D. Grider, and P. Bauhahn. 1991. Refractory self-aligned-gate GaAs FET based circuit technology for high ambient temperatures. Pp. 523-528 in Transactions of the First International High-Temperature Electronics Conference, Albuquerque, New Mexico.

Sze, S.M. 1981. Physics of Semiconductor Devices, 2nd ed. New York: John Wiley & Sons.

Zipperian, T.E. 1986. A survey of materials and device technologies for high temperature (T>300 °C), power semiconductor electronics. Pp. 353-365 in Proceedings of Power Conversion Intelligence, October.

Appendix C: High-Temperature Microwave Devices

Semiconductor devices are making an ever-greater impact in system applications as they are increasingly utilized for microwave and radio frequency (RF) power generation and amplification. The history of this impact can be traced back to the early 1970s when silicon diodes were developed for microwave detection and power generation. The utility of the diodes extended to frequencies unattainable by silicon bipolar transistors that were limited by their long charging-time constants and low carrier-diffusion rates through the base layer. Silicon FETs were developed for digital applications, but their potential for microwave use was pre-empted by improvements made in gallium arsenide (GaAs) materials technology. GaAs offered high electron mobility and high saturated velocity not available with silicon and held the promise of much better microwave performance. The first GaAs devices were Gunn diodes that used the negative differential mobility available in the material. Throughout the early 1980s there was steady progress in improving the quality of GaAs wafers and devices made from them.

The premier device of interest was the GaAs MESFET. Development benefited from electron beam gate writing technology established during the same time period. Gate lengths achievable for GaAs MESFETs decreased to submicron dimensions, enabling a number of high-performance applications through the millimeter-wave bands. At the same time, GaAs IMPATT diodes set records for power output. Such diodes are now combined to form units that can achieve the kilowatt level at frequencies above 30 GHz. Indeed, GaAs IMPATT diodes are a serious contender for applications now requiring vacuum tubes.

In the late 1980s and early 1990s, materials technology again advanced with the widespread application of MBE and metal-organic chemical vapor deposition (MOCVD) techniques. Heterojunction structures made from the III-V compound semiconductors (Ga,Al,In)-(As,P) could be fabricated resulting in material combinations in which band structures were engineered to optimize device performance. High Electron Mobility Transistors (HEMTs) and Heterojunction Bipolar Transistors (HBTs) improved on the performance of microwave devices.

Concurrent with the development of improved III-V microwave devices was the development of monolithic circuit technology. For this technology, circuits necessary to tune transistors electrically and combine their power were printed directly on the same semi-insulating GaAs substrates incorporating the active devices themselves. The result was a cost and weight savings in addition to considerably enhanced functionality of the chips over discrete devices. Materials technology was again a key to the progress of the field. For integrated circuits to be cost effective, large defect-free wafers were required. These wafers were developed as the industry recognized the need for them. Development work was aided by the commercial demand for LEDs requiring GaAs substrates.

In recent years, devices on indium phosphide (InP) substrates have come under development. HEMTs and HBTs on InP can have performance advantages over devices designed on GaAs substrates. This is due to the broad range of materials which can be grown hetero-epitaxially on InP. Also, compared to GaAs substrates, InP offers higher avalanche breakdown fields and higher thermal conductivity (e.g., the thermal conductivity of InP is 0.7 W/cm·K; GaAs is 0.54 W/cm·K). Nevertheless, microwave devices on GaAs provide the benchmarks of performance against which other materials are judged.

As systems designers consider the use of solid-state microwave devices, they are confronted with the need for maintaining device temperatures low enough to ensure efficient operation and provide for reliability. In general, operating temperatures of GaAs transistors and diodes must be held below 150-200 °C. Above these temperatures, parasitic resistance increases and saturated carrier velocities decrease, both contributing to degradations in gain performance. Furthermore, there is a significant increase in the pace of deterioration in the contacts—and in some cases the doped layers—due to elevated diffusion coefficients of particular dopants such as zinc (p-type in GaAs). Deterioration is at least partially thermally activated, with applied voltages and currents also playing a role.

The major implications of the need to limit the temperature of microwave devices are twofold: (1) power density must be limited at the active layers since much of the temperature rise is due to heat-spreading resistance (known as thermal resistance) through the substrate very near to the active regions, and (2) attention must be paid to cooling the chip mount to minimize ambient temperature rise due to the time average power dissipated. The second item has a major impact on the utility of GaAs-based devices in microwave systems. Cooling systems must be provided to remove heat from the system and maintain heat-sink temperatures at acceptably low levels. The operating temperature of the device is given by Equation (C.1)

$$T_j = T_o + R_{th}P_{dis} , \qquad (C.1)$$

where T_j is the active-layer temperature, T_o is the ambient heat-sink temperature, R_{th} is the device thermal resistance, and P_{dis} is the dissipated power. It is evident from Equation (C.1) that higher permissible operating temperatures and lower thermal resistance will allow higher ambient temperatures in systems. This will have a favorable impact on size, weight, complexity, and cost. Furthermore, if higher operating temperatures could be utilized, circuits with higher power-dissipation densities would be allowed. For low thermal resistance, a high thermal conductivity material is required since, as noted above, most of the thermal resistance is provided very near to the active region of the device where the heat has not yet spread over a large area.

Higher permissible power dissipation is an advantage to the system designer only if it adds functionality to the circuit (by allowing more transistors per unit area) or if it relates directly to improved power performance. This point is illustrated by expressing the dissipated power, P_{dis}, in a transistor in terms of its gain, G, and its RF output power, P_{out}:

$$P_{dis} = [(1/h_D) + (1/G) - 1]P_{out} , \qquad (C.2)$$

where P_{DC} is the DC input power to the transistor and $h_D = P_{out}/P_{DC}$ and h_D is called the drain or collector efficiency depending on the transistor type. A useful parameter is the power-added efficiency (PAE) defined by

$$PAE = (P_{out} - P_{in})/P_{dc} = h_D[1 - (1/G)] . \qquad (C.3)$$

Normalized curves of power dissipation are shown in Figure C-1 indicating sensitivity to both gain and efficiency. If a transistor with a given thermal resistance can withstand higher operating temperature but is less efficient, the ambient temperature must be reduced correspondingly and the advantage is lost. Furthermore, the prime power required by the transistor is increased, making it less suitable for use. An alternative is to reduce the output power required from the transistor. This, too, has undesirable system implications.

In summary, good performance (gain, efficiency, and power output) is required from any microwave semiconductor device intended for operation at high ambient temperature. Only in rare cases will poorly performing devices be used simply to withstand high ambient temperature.

The semiconducting materials discussed in this report, SiC, diamond, and III-V nitrides are candidates for use at

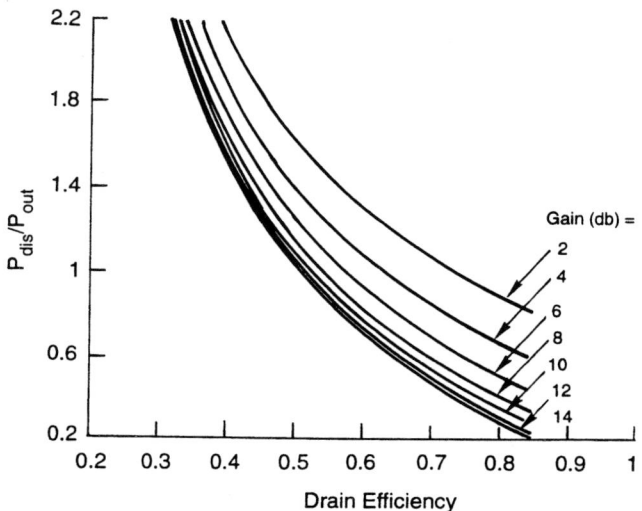

FIGURE C-1 Contours of normalized power dissipation on the gain-efficiency plane.

high temperatures according to the criteria discussed. These candidate materials can be classified as "emerging materials," as GaAs was an emerging material of the early 1970s. Thus, in assessing the potential for microwave devices constructed from SiC, diamond, and nitride materials, this appendix separates the "fundamental" properties inherent to these materials from those related to technology deficiencies that will almost certainly be overcome in the long run. This fundamental category includes such parameters as mobility, saturated velocity, dielectric constant, and avalanche breakdown electric field.

Two often-cited "figures of merit" are due to Johnson (1965; JFM) and Keyes (1972; KFM). JFM accounts for the fact that in an intrinsic device (i.e., one without parasitic resistance or reactance) there is a tradeoff between the time a carrier spends gaining energy in an electric field as it drifts through a device and the response time of the device. JFM is related to electronic properties and does not account for thermal effects. JFM also does not predict the ultimate device active volume, proportional to power.

A better figure of merit is KFM. Impedance considerations and the resulting increase in thermal resistance as devices are made smaller are accounted for by use of KFM. Keyes assumes that smaller devices are inherently faster in response at a fixed input impedance level. But devices cannot be made smaller without increasing the thermal resistance and thereby limiting the power output. This introduces the thermal conductivity as a factor. The breakdown field is not significant in this figure of merit since KFM addresses a thermal rather than an electronic limit.

Calculated JFM and KFM for a variety of materials are shown in Chapter 3 in Table 3-1. For JFM, the high breakdown field dominates, making all of the wide bandgap materials attractive when compared to silicon, germanium, and GaAs. There appears to be no significant difference in the predicted merit among the various wide bandgap materials. The high thermal conductivities of the wide bandgap materials increase the values of KFM. Also, their lower dielectric constants reduce the capacitance per unit area thereby further increasing KFM.

The figures of merit suggest that electronically limited devices such as FETs should have higher power density in the wide bandgap materials. For thermally limited devices such as bipolar transistors or IMPATTs, higher power density should also be achievable in the wide bandgap devices. These predictions must be moderated by the fact that both figures of merit give a very crude picture of the situation since only the "intrinsic" device is considered. The figures of merit do not account for parasitic resistance and other detailed effects that limit gain and efficiency. The requirements of matching devices operating under bias conditions that take advantage of the material properties is also omitted from the figure-of-merit analysis. For example, is it possible to operate devices at high voltages when the necessary high impedance loads are in conflict with conjugate matching requirements for gain and efficiency? These considerations are addressed in more detail in the next section.

The figures of merit just discussed are based on some of the fundamental properties of the materials. Ultimately, technology issues must be confronted. Such issues include wafer size, defect density, contact resistance, and stability of contacts at elevated temperature. Fundamental properties and technology issues are not totally unrelated, and some important factors fall between the two categories. For example, sheet resistivity and contact resistance may both be related to the alloy method and the metallurgy used to form the contacts. The choice of doping method (e.g., growth versus implantation) may relate mobility (fundamental) and doping density (technology). These relations can only be resolved in the course of time as materials, technology, and devices are developed and tested experimentally.

It is difficult to make predictions of ultimate performance based on today's demonstrated technology. Instead, this appendix considers the various devices under development to highlight the relative importance to performance of physical parameters. The numbers used in the analysis are based on present-day state of the art. Since the committee is working from an incomplete knowledge of material physical parameters as a function of temperature and doping, some assumptions are made in the analysis. It is hoped that these assumptions represent the correct order of magnitude for devices that can ultimately be built from wide bandgap materials. From the analysis, it should be evident which technology elements must be given the most attention (for example, contact resistance), and this should be taken as motivation for further research and development.

There are two general approaches taken to make predictions of microwave performance: detailed numerical computer models and analytic approaches. Numerical models can yield highly self-consistent results that may not represent the best global operating conditions even when optimization is attempted. The details may be inaccurate due to imperfect knowledge of the physical parameters or circuit constraints. Furthermore, such an analysis often does not clearly expose the physically important parameters. Therefore, this appendix applies the analytical approach to the most important microwave devices.

BASIC DEVICE TYPES

The basic microwave device types for high-temperature semiconductors are the same as those now commonly made from GaAs and silicon. Candidates are the following:

(1) bipolar junction transistors (BJTs),
(2) static induction transistors (SITs),
(3) junction field effect transistors (JFETs),
(4) metal-semiconductor field effect transistors (MESFETs),
(5) heterojunction transistors (HJTs),
(6) impact avalanche transit time (IMPATT) diodes.

Excluded from this list are both enhancement- and depletion-mode MOSFETs (see Figure C-2), which are expected to have limited utility at microwave frequency when compared with MESFETs. Oxides tend to trap charges within them and also at the interface between the oxide and semiconductor. This condition results in the screening of the conducting channel to high-frequency excitation when the transistor is not driven into saturation as it would be in switching applications. Furthermore, MOSFETs can have poor electron surface mobility due to roughness at the oxide-semiconductor interface. If p-channel MOSFETs are used in the enhancement mode, channel access must be made by holes that have low mobility or a high parasitic resistance will result.

Despite this pessimism for the potential of wide bandgap MOSFETs for microwave application, it should be noted that MOSFETs fabricated from both 6H- and 3C-SiC have shown good DC properties, even at

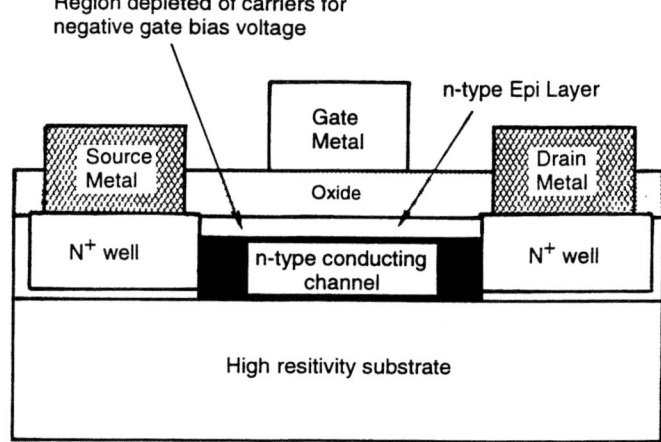

FIGURE C-2 Enhancement- and depletion-mode MOSFETs.

temperatures as high as 400 °C. It would be worthwhile exploring the frequency response of such devices to determine in more detail their limits of performance. Indeed, MOSFETs may have a useful role to play in low-frequency operation where the reliability at high temperatures is a primary consideration. The oxide-metal interface may be more stable with temperature than that formed between the metal and the semiconductor. Changes in that interface during long-term operation may have a less significant effect on device performance than one might encounter in a MESFET.

Bipolar Junction Transistors

The basic structure of the bipolar junction transistor (BJT) is shown in Figure C-3. A voltage on the base is

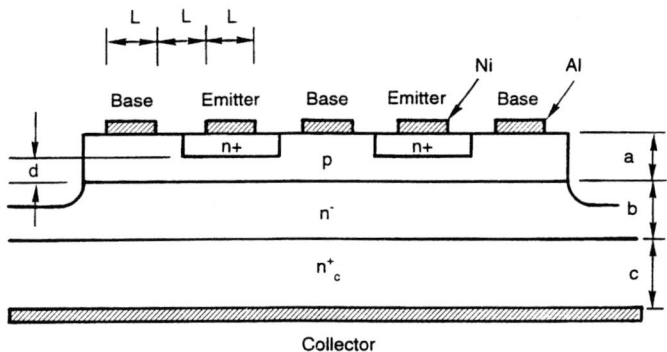

FIGURE C-3 Structure of a bipolar junction transistor. SOURCE: Trew et al. (1991), © 1991 IEEE.

used to forward-bias the base-emitter junction and charge diffuses across this junction to the reverse-biased collector-base junction. In the base, carriers are subject to bulk recombination and have a lifetime on the order of 1-10 nanoseconds for GaAs, 3C-SiC, 6H-SiC, diamond, and GaN but not for silicon where it can be in the millisecond range due to the indirect bandgap and the low density of recombination centers. Other recombination mechanisms that depend on surface treatment are operative at the edges of the base and emitter regions, particularly at low current densities. It is expected that these effects will be related to device fabrication details and there is no proven model for them at present.

The charge moving across the base of a BJT that does not recombine on its journey is accelerated across the reverse-biased collector layer delivering power to the external load. An n-p-n transistor is preferred since electron mobility, and hence diffusivity in the base, is higher than that of holes. Higher diffusivity results in higher minority carrier-density gradient in the base and therefore higher current gain. A penalty paid for this is that the extrinsic base resistance is higher in an n-p-n transistor than for a p-n-p transistor due to the lower hole mobility.

There are several examples of SiC BJTs that have shown reasonable DC current versus voltage curves up to 400 °C (Palmour et al., 1993). In early work (Muench et al., 1977), n-p-n devices were fabricated from CVD layers and had DC current gain in the 4-8 range and leakage current of 10^{-5} A/cm^2 at V_{ce} = 40 V. Although demonstrations of DC current gain is a necessary first step for a viable microwave device, the issue of frequency response is not addressed in such measurements.

The frequency response of a transistor is comprised of an "intrinsic" time related to the rearrangement of charge in the basic device layers and an "RC" time constant related to charging the capacitance parasitic to the transistor through parasitic resistance of the layers and contacts. Until the parasitic capacitances are fully charged, the voltages at the intrinsic device layers will not reach steady state. The intrinsic response time is characterized by the frequency at which the current gain becomes unity, f_t, as

$$t = 1/(2\pi f_t) . \qquad (C.4)$$

A second characteristic frequency is f_{max}, the frequency at which the unilateral gain becomes unity according to Equation (C.5):

$$U = (f_{max}/f)^2 , \qquad (C.5)$$

and it can be shown that f_{max} is given approximately by Equation (C.6):

$$f_{max} = [f_t/8\pi(RC)]^{1/2} , \qquad (C.6)$$

where R is the extrinsic base resistance and C is the parasitic base-collector capacitance. Equation (C.5) and (C.6) illustrate the importance of parasitic resistance and capacitance in determining the gain of the transistor. To reduce C, the base contacts must be small, but this, in turn, increases base contact resistance that is already high due to the low base doping required for DC current gain. The low base doping also increases the sheet resistance under the base contact and between the base contact and emitter contact. These effects severely reduce the frequency response of silicon and GaAs BJTs rendering them essentially useless above 3 GHz.

Parasitic effects would have an even more detrimental effect on bipolar transistors constructed of wide bandgap semiconductors such as SiC and GaN. Bipolar transistors fabricated from diamond cannot yet be contemplated since no effective n-type dopant has been found for the material.

More detailed simulations have been conducted of BJTs predicting that performance would be severely limited by ohmic contact resistance now achievable and by high base sheet resistance due to low hole concentrations necessary (Gao et al., 1994). The base resistance would further be increased by the need to use narrow base widths (less than 500 Å) to allow rapid base transit of electrons. Other simulations of SiC BJTs were conducted by Trew et. al. (1991) who showed that performance falls rapidly above 1.5 GHz (Figure C-4). Better performance is predicted from a 6H/3C-SiC HBT where a valence-band discontinuity between the two polytypes could confine holes to the base under forward bias (Gao et al., 1994). Thus, the base carrier concentration could be increased by several orders of magnitude to up to 2×10^{19} cm^{-3}. In the analysis, base doping was assumed to be graded over the layer by two orders of magnitude to provide a built-in electric field and thereby speed transport of electrons across the base layer. Also assumed was a base contact resistance of 10^{-6} $\Omega \cdot$cm^2, which is a somewhat optimistic value for p-type material. The results suggest that the devices at 1 GHz would be usable up to 450 °C (Gao et al., 1994). If base contact resistance were higher, the performance would not be acceptable.

Little can be said about the potential of the other wide bandgap materials for BJT application, but the situation is likely to be similar to that of SiC. An interesting possibility would be the GaAlN/GaN heterojunction bipolar transistor, especially if constructed on a SiC substrate with high thermal conductivity. GaAlN and GaN are lattice matched over a wide aluminum concentration and the bandgap can be varied with an aluminum mole fraction between 3.5 eV and 6.5 eV. Such a heterostructure would be useful if a significant valence-band discontinuity could be established without an excessive large conduction-band discontinuity. A discontinuity in the conduction band would lead to an energy barrier to electron injection from the emitter into the base. There would be a large collector voltage threshold voltage for turning on the transistor. Low hole mobility (≈ 50 cm^2/V·s) and an associated high parasitic base resistance is a potential difficulty with GaAlN/GaN heterojunction bipolar transistors. More detailed information about the properties of GaAlN/GaN combination should be obtained experimentally to asses the potential of such devices for heterojunction bipolar use.

Static Induction Transistors

The static induction transistor (SIT) is a three-terminal device that has several features making it attractive for development using wide bandgap material. The structure is shown schematically in Figure C-5. A source contact is positioned on the surface of an n-type crystal between two p-type regions. The p-n junction forms a gate that, when reverse-biased, can regulate the flow of electrons from the source by raising and lowering the potential barrier seen by the carriers. This modulates the number of carriers

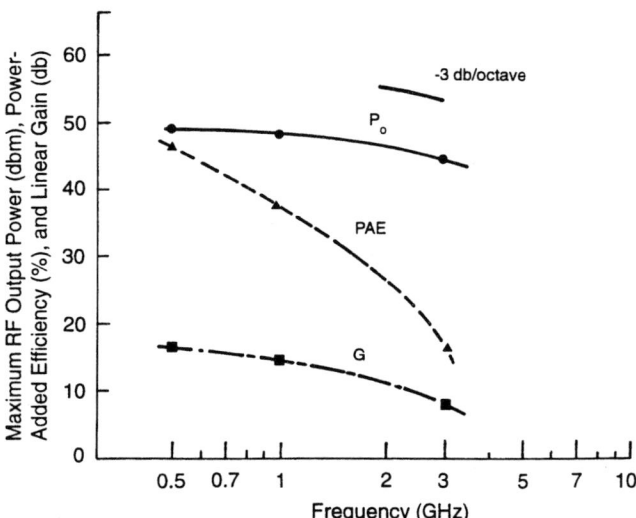

FIGURE C-4 Simulated microwave performance of SiC BJTs. SOURCE: Trew et al. (1991), © 1991 IEEE.

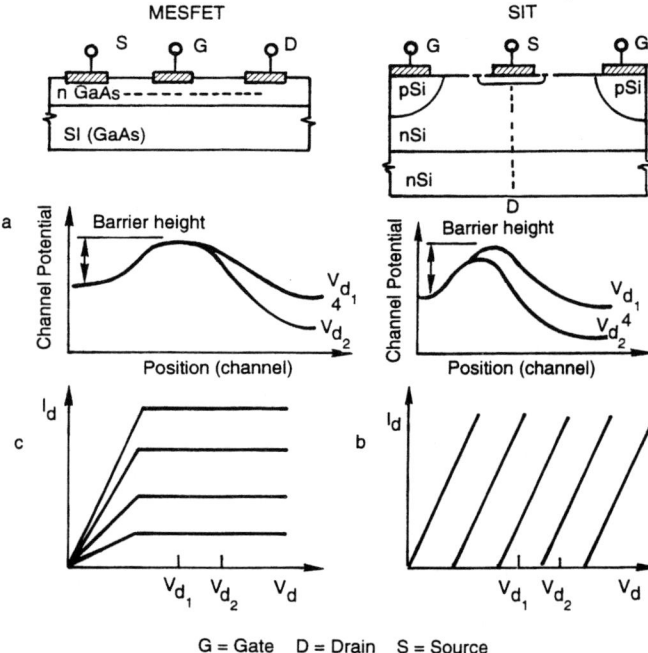

FIGURE C-5 Comparison of SIT with MESFET: (a) potential gate barriers established, (b) resulting current-voltage curves for SIT, (c) generic MESFET I-V curves. SOURCE: Trew et al. (1991), © 1991 IEEE.

collected at the drain, which is heavily doped n-type. The reverse bias established between the gate and the drain provides an electric field to accelerate the injected carriers. The potential barrier established by the gate and the resulting current-voltage curves are shown in Figure C-5 where they are compared with the corresponding curves for MESFETs. It is noted that the barrier height is a strong function of the applied drain voltage and that this results in a high output conductance in the I-V curves of the SIT. The current-voltage contours of an SIT resemble those obtained from a vacuum-tube triode.

There is reason to believe that SITs from wide bandgap material, and particularly in SiC, can be useful for high-power application. The structure uses a vertical design in which carriers move in a direction perpendicularly to the crystal surface. This eliminates the requirement of submicron gate fabrication for high-frequency operation. Transit time is determined by drain-layer thickness that can be controlled and made very small during epitaxial growth. For a vertical device, there is a tradeoff between breakdown voltage and transit time. This is predicted by the JFM to be favorable for SiC compared with silicon. The devices can be operated at high drain voltages ($V_D = 100$ V) for thick, lightly doped channels, provided that transit time through the channel does not limit frequency. In contrast to the MESFET, the current density through a SIT channel is not limited by its thickness and can be much higher when normalized to the total length of the source contact. Current density for the SIT is most properly normalized to the source contact area and described in A/cm^2 rather than mA/mm. The SIT channel is depleted of carriers and has a high electric field making it more immune to space charge effects that eventually limit the current density.

SITs can be fabricated on highly doped substrates. For low microwave frequencies where hybrids can be used, it is not necessary to use semi-insulating material as with the MESFET. Semi-insulating substrates would be necessary only if monolithic circuits are desired. Using conductive substrates would permit a more robust materials technology and allow a wider range of growth and bonding options. Frequency response could be improved since traps and defects in the substrates would be less important.

The SIT is a thermally limited device. The high thermal conductivity of SiC gives it an advantage over silicon and GaAs for SITs predicted by the KFM. The lower dielectric constant of SiC reduces the output capacitance per unit area of the device allowing larger devices. Also decreased is the parasitic source to gate capacitance that provides degenerative feedback and reduces gain in a manner similar to the case of bipolar transistors described earlier. The thermal and electrical advantages noted for SiC should be even more pronounced if diamond could be used for the SIT. The diamond application would require a suitable n-type dopant.

Having considered the attractive attributes of SITs, some potential problems should be noted. The gain available is a strong function of the parasitic gate resistance. It is estimated that gate contact resistivity would have to be less than 1×10^{-5} $\Omega \cdot$cm^2 to produce usable gain above 2 GHz. Compared to the parasitic base resistance of homojunction bipolar transistors, parasitic contact resistance can be reduced since base doping can be much higher. The base layer could be produced by ion implantation of p-type dopant. Alternatively, the p-type gate could be replaced by a Schottky contact. A problem with this approach is that sidewall metallization of the channel layer would be needed to control the potential effectively; it would be difficult to control substantial charge flow using the edge fields generated by a totally horizontal gate contact.

An important issue affecting performance is the tradeoff between output-matching requirements for power and maximum available gain. This problem is an important one and is treated in more detail in the section below. Simply put, conjugate match at the output is required for maximum gain, while a different load impedance is most often required for maximum power. The output conductance of a typical SiC SIT as determined from calculated current-voltage curves is around 460 $\Omega \cdot \mu$m normalized to the total source length. It is predicted that SiC SITs can give up to 1.8 W/mm power density at 3 GHz at $V_D = 100$ V. The necessary current density would then be 1.8 mA/μm and the load impedance needed to sustain this voltage would be around 700 $\Omega \cdot \mu$m. The difference in the output-matching requirements for power and gain will mean that a compromise output impedance must be found to optimize performance. It will be problematic to relinquish gain in this compromise; gain will already be at a premium for the reasons discussed above and will be further reduced from small-signal values due to nonlinearity of the SIT and resulting gain compression. Accord-

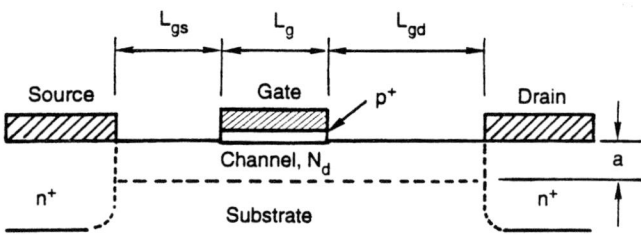

FIGURE C-6 Structure of the Junction Field Effect Transistor (JFET). SOURCE: Clarke et al. (1993), Courtesy Westinghouse, Inc.

ingly, it may turn out in practice that operating voltage of low-frequency SITs in wide bandgap semiconductors will be well below half the drain breakdown voltage. In contrast, the higher output impedance of FETs allow closer values of the gain and power impedance levels.

Junction Field Effect Transistors

In a junction field effect transistor (JFET; Figure C-6), motion of charge is controlled by a junction gate placed on the surface of the crystal in a space separating the source and drain contacts. The conducting channel is an n-type layer with doping in the 10^{17} cm^{-3} range. Relative to the source, the drain is positively biased and the p-n junction forming the gate is reverse-biased. When the gate voltage is modulated by an external signal, the thickness of the conducting channel varies, thereby controlling the current flow to the drain. Power gain is obtained. Typical current-voltage curves are shown in Figure C-7 for several temperatures.

The JFET is similar to the MOSFET operating in the depletion mode. Enhancement-mode operation is not possible in the JFET since the gate would be forward-biased and would draw current, increasing power dissipation in the gate and possibly introducing minority charge storage effects that would slow the frequency response of the transistor. An advantage of the JFET compared with the MOSFET is that surface charge at the gate junction is absent. Current moves in a channel region away from the possibility of surface scattering. A disadvantage in most semiconductors is that the low mobility of holes adds a parasitic gate resistance that can reduce gain. JFETs have an advantage over MESFETs in that the gate metal is further from the active channel. At high temperature and RF power the gate metal can diffuse into the semiconductor and degrade the gate. JFETs could be more reliable.

The device structure shown in Figure C-6 requires that a semi-insulating substrate be positioned below the gate to limit channel thickness. This can be a problem in wide bandgap semiconductors that do not have high resistivity substrates available. An alternative is shown in Figure C-8 where the gate is an n-type epitaxial layer placed below the channel. The channel is etched to a desired thickness and passivated with a surface oxide. This oxide must be high quality and, like the MOSFET, cannot contain traps or extraneous charge that might cause backgating effects. Furthermore, the interface between the oxide and the semiconductor must be free of scattering sites that would reduce the mobility of carriers.

Inverted JFETs have been fabricated (Kelner et al., 1987, 1989) from 3C-SiC on a 6H-SiC substrate. DC transconductance as high as 20 Ms/mm was obtained, but the devices could not be pinched off (the drain current was brought to zero by reverse-bias gate voltage). The transistors exhibited a high output conductance. In another experiment (Kelner et al., 1991), 6H-SiC homo-epitaxial transistors gave DC transconductance up to 17 mS/mm and could be completely pinched off with -40 V applied to the gate. DC transconductance decreased to 1.7 mS/mm at elevated temperature (400 °C) due to decreasing mobility of electrons in the channel.

One problem with inverted JFETs in wide bandgap semiconductors is the presence of high built-in voltage (approximately equal to the bandgap). The built-in voltage partially depletes the conducting channel even with no applied gate voltage. For a given channel thickness, this lowers the maximum current available and reduces the effectiveness of short gates in pinching off the channel. Another disadvantage of the inverted structure results from the extension of the gate layer under the source and drain contacts. There is a resulting significant increase in gate capacitance, C_{gs}. This can severely reduce gain for a given transconductance, g_m, since f_t is inversely proportional to C_{gs}.

Metal-Semiconductor Field Effect Transistors

Relatively more attention has been paid to MESFETs than any other microwave device in wide bandgap semiconductors. Goals for MESFETs under development include producing X-band power density three times that of GaAs with power-added efficiency twice that of GaAs. DC operation of MESFETs has been demonstrated by

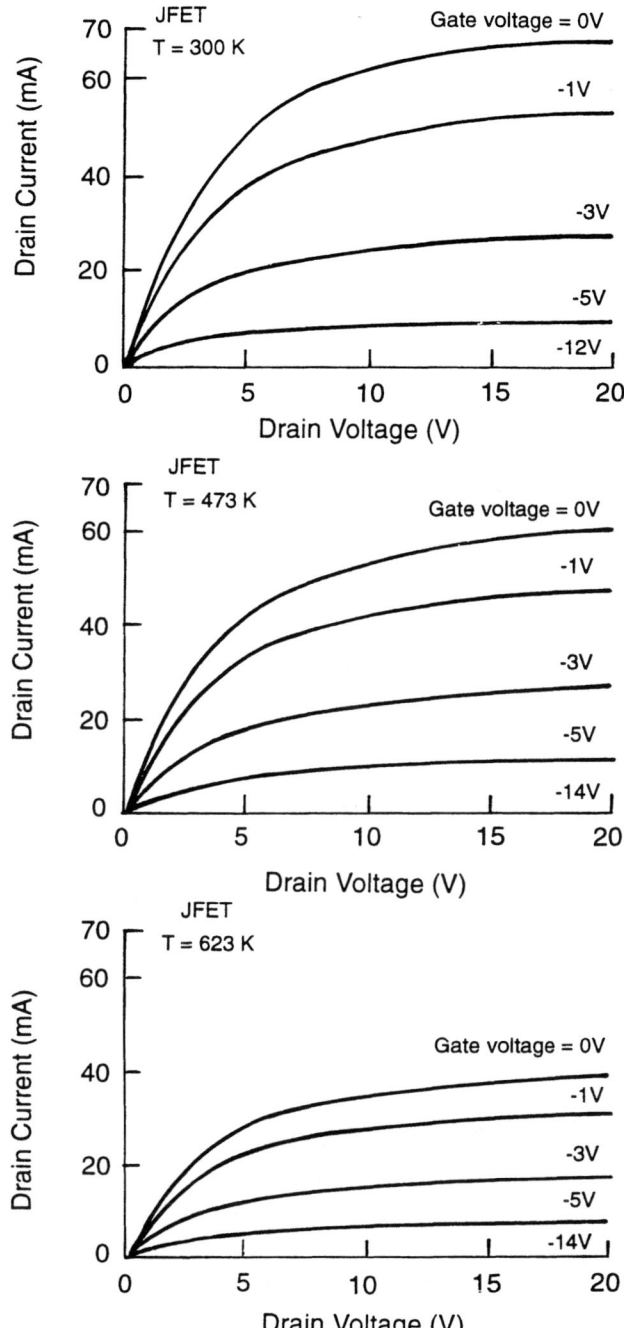

FIGURE C-7 Typical current-voltage curves for a JFET at various temperatures. SOURCE: Palmour (1993), Courtesy of Cree Research, Inc.

SiC. Typical measured RF results are shown in Figure C-9.

Measurements of power performance have been conducted by Westinghouse and Cree Research Incorporated. At Westinghouse, MESFETs with frequency-of-unity maximum available gain of 5 GHz gave 2 W/mm at an operating frequency of 1 GHz. Total output power was 1 W with a drain bias voltage of 75 V. At Cree Research Incorporated, 0.6 micron gate-length MESFETs of 6H-SiC had f_t = 5.5 GHz and unilateral f_{max} = 3.8 GHz. For this device, V_{ds} = 45 V and I_d = 126 mA; device gate width was 340 μm. Slightly better performance was obtained from MESFETs on 4H-SiC for which f_{max} = 5.8 GHz at V_{ds} = 20 V. One might expect that the lower operating voltage in the 4H case would result in increased f_t and that this would in turn result in higher measured f_{max}. It was found that f_t was actually reduced, however. This indicates that the improvement in f_{max} might be attributed to the higher mobility of the 4H-polytype and the resulting lower parasitic resistance.

Impact Avalanche Transit-Time Diodes

GaAs IMPATT diodes have been developed in the last few years, emerging as the solid-state device giving the highest power in the microwave and millimeter-wave frequency bands (Figure C-10). The wide bandgap semiconductors have the potential to give higher power density and to operate at higher temperatures. To achieve a better view of this potential, this section first describes the general features of the IMPATT diode and its operation.

Several material structures are possible for IMPATT diodes and are shown in Figure C-11. In its simplest form

many researchers over a range of temperatures up to 300 °C. Table C-1 summarizes room-temperature DC gain for various FETs of SiC. Operation of MESFETs has been simulated and samples measured for small-signal gain up to 10 GHz. Submicron gates have been used for

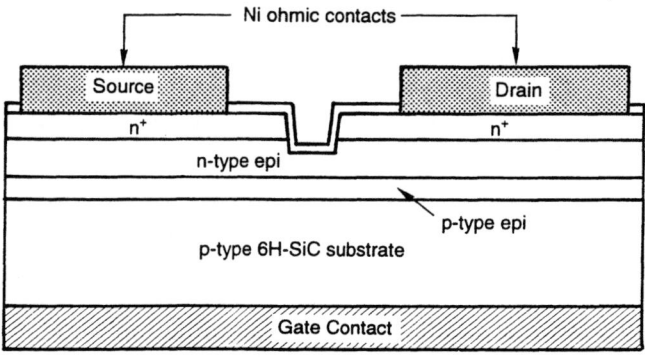

FIGURE C-8 Structure of an inverted JFET in SiC. SOURCE: Palmour (1993), Courtesy Cree Research Inc.

TABLE C-1 Summary of Room-Temperature DC Gain for Various Field Effect Transistors of SiC

Material	Device/Mode	Gate Length (μm)	g_m (mS/mm)	Channel Mobility (cm_2/V·s)	Highest Operating Temperature (°C)
3C-SiC	MOSFET/Enhancement	5	0.46	Low	650
6H-SiC	MOSFET/Enhancement	7	2.8	46	450
3C-SiC	MOSFET/Depletion	2.4	10	37	750
6H-SiC	MOSFET/Depletion	5	2.3	21	450
6H-SiC	MESFET	24	4.3	300	450
6H-SiC	MESFET	0.4	30	—	23
3C-SiC	JFET	4	20	560	23
6H-SiC	JFET	5	17-20	250	627

SOURCE: Morkoc et al. (1994).

(the single-drift diode), the structure (Masse et al., 1985) consists of a p$^+$-n junction with an n$^+$ contact. Under reverse DC bias, the electric field profile of Figure C-11 is established with the junction electric field sufficiently high to cause impact avalanche breakdown in the region. The electrons generated at the junction drift toward the n$^+$ contact. If the current is limited by a high impedance bias circuit, a steady current at the breakdown voltage is maintained. If, in addition, there is an RF voltage applied, the peak voltage will generate a charge clump that moves across the n-type drift zone. The current induced in the external RF load at the fundamental frequency will be 180 degrees out of phase with the RF voltage. Thus, an effective negative resistance is established across the diode terminals at the operating frequency. In general, the drift region length scales inversely with operating frequency in order to maintain the proper current–voltage phase delay.

A simplified, equivalent circuit for an IMPATT diode embedded in a microwave circuit is shown in Figure C-12. The capacitance results from the depleted drift region and the external RF circuit provides the inductance and the RF load. A parasitic series resistance is included. When the circuit is properly tuned, the reactances cancel and a self-sustaining RF current builds in the loop formed by the device and RF load. As the current grows, the diode negative conductance decreases and voltage reaches a steady state when the net resistance around the diode circuit loop is zero. This is the condition for oscillation. If the load resistance is increased to a point where oscillations do not occur, the diode acts as a reflection amplifier.

The parasitic series resistance depicted in Figure C-12 limits the device area and power. Compared to values in GaAs, saturation resistance, R_s, is expected to be much larger in wide bandgap materials because of generally lower mobilities and higher contact resistance. In terms of the circuit parameters defined in Figure C-12, the RF power output of the IMPATT is given by

$$P_{out} = (1/2)(|G| - |B|^2 R_s)V_{rf}^2 , \qquad (C.7)$$

where $|B| = 2\pi feA/X_D$, f is the frequency of operation, e is the dielectric constant, A is the device active area, and X_D is the total thickness of the depleted active layer (Adlerstein and Moore, 1981). The voltage amplitude at the transistors terminals is V_{rf}. The first term in Equation (C.7) represents the power available from the intrinsic device and the second term is due to I^2R losses in the parasitic series resistance. The device admittance at constant current density and susceptance are proportional to junction area while the series resistance is at best inversely proportional to area, decreasing more slowly than 1/A in most cases. Thus, as device area increases, output power passes through a maximum, eventually declining; it is important that R_s not be too large or negative resistance will not be obtained. High R_s has an increasingly damaging effect as frequency is increased. The susceptibility due to the output conductance is proportional to frequency with $B = 2\pi f C_d$, where C_d is the diode capacitance. The reduction of net negative conductance is proportional to B^2 producing a strong frequency roll-off in power for an IMPATT of a given area. The problem of high R_s is compounded at the large V_{rf} levels required for high power or saturated oscillator operation since $|G|$ generally decreases with increasing V_{rf}.

High parasitic resistance is a particularly troublesome, potential problem for wide bandgap semiconductors. In the past, typical specific contact resistance for SiC n$^+$ material of 10^{18} doping range is around 1×10^{-4} $\Omega \cdot cm^2$ and about 5×10^{-4} $\Omega \cdot cm^2$ for p$^+$ at the same doping. Contact resistivity on n$^+$ material is relatively insensitive to temperature up to 400 °C, while resistivity on p$^+$ layers decreases significantly due to increasing carrier density. Nevertheless such resistivities are two orders of magnitude greater than for GaAs. Recently, researchers have reported contacts with an order-of-magnitude lower resistivity for both p$^+$ and n$^+$ contact types. Efforts should continue to further reduce the contact resistance.

A more fundamental issue is the low carrier mobility of both electrons and holes in SiC. Experimental and theoretical study will be required to determine the best combination of material and contact parameters for best performance.

At a given frequency, power density of an IMPATT diode is determined by its operating voltage (roughly the same as its reverse breakdown voltage corrected for space-charge effects) and the maximum temperature that can be tolerated at the junction without performance degradation or reliability problems. The maximum operating temperature usually dictates the upper bound on current in continuous-wave operation. When the diodes are operated in the pulsed mode, power density can be considerably higher since heating is truncated at the conclusion of the pulse and the IMPATT has an opportunity to cool between pulses. For very short pulses, a current density limit is eventually reached where space-charge effects degrade the performance. Power limitations in the pulsed mode can be overcome by using more sophisticated doping profile designs.

One such design of particular importance for silicon and GaAs is double-drift Read structure (Figure C-11). It is worthwhile to inquire if IMPATTs of wide bandgap semiconductors would benefit from such designs. In a double-drift Read diode, there is a drift region for the holes as well as for the electrons. There is also a doping spike on either side of the junction that abruptly decreases the electric field in the drift region to confine the avalanche zone to the region near the junction. This results in increased efficiency since a higher fraction of the applied bias voltage is used to pull charge through the drift zones. The operating voltage of a double-drift diode

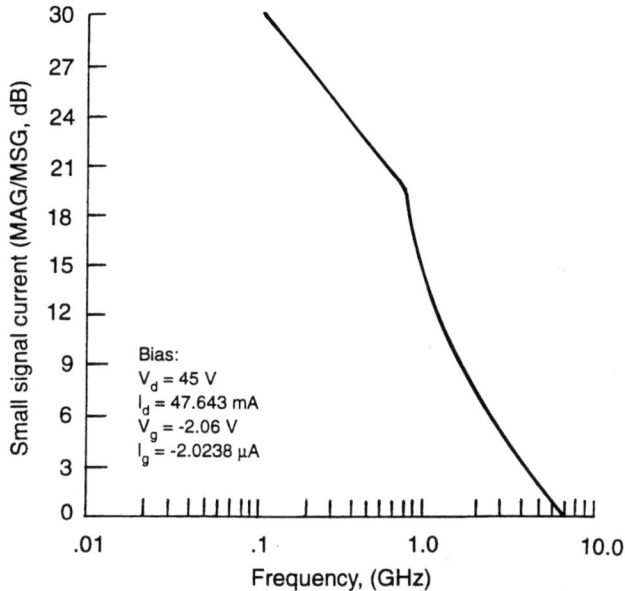

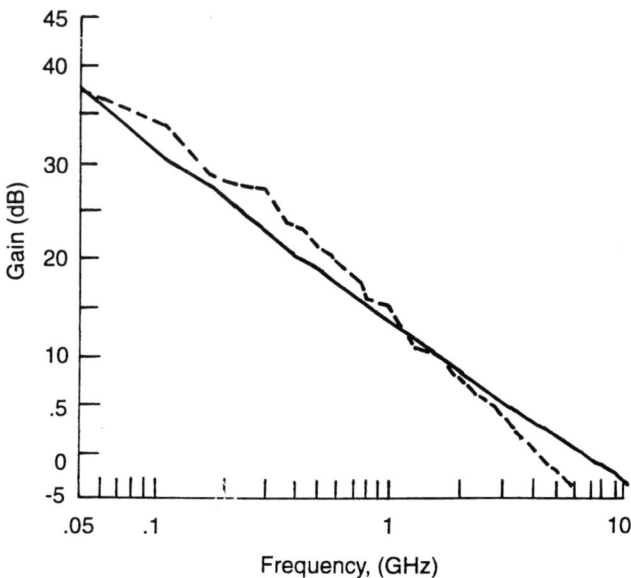

FIGURE C-9 Measured small-signal current and unilateral gain for SiC MESFETs. SOURCE: (a) Hobgood (1993), Courtesy Westinghouse; (b) Palmour (1993), Courtesy Cree Research, Inc.

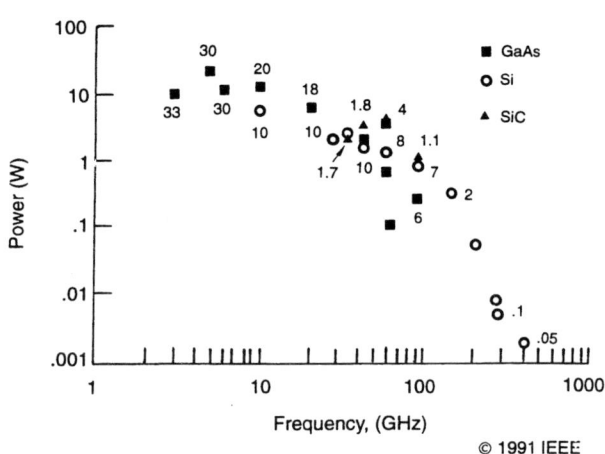

FIGURE C-10 IMPATT diode performance compared with projections for wide bandgap semiconductors. Values at points are DC-RF power conversion efficiencies. SOURCE: Trew et al. (1991), © 1991 IEEE.

is roughly 60 percent higher than for its single-drift counterpart. Accordingly, double-drift IMPATTs have a higher power density for a given current density. Furthermore, the two drift regions act as capacitors in series, resulting in a decrease in junction capacitance per unit area by nearly a factor of two. Thus, device area can be doubled with a resulting increase in power at a given power density.

Given these basic elements for the operation of IMPATT diodes, and ignoring potentially high parasitic series resistance, the wide bandgap semiconductors provide interesting possibilities. SiC has potential for high power density in IMPATT operation due to the high breakdown voltages that can be achieved. Electron drift regions for SiC could be longer for a given frequency than for GaAs due to the higher saturated velocity (2×10^7 cm/s) in SiC. Of course, it would be necessary for the higher velocity to persist at elevated temperatures. There would be a decrease in the capacitance per unit area resulting from the longer drift region, and this capacitance would be further decreased by the lower dielectric constant of SiC. Larger-area diodes would be possible.

Double-drift IMPATT diodes could be fabricated using SiC, but these diodes would not have equal thickness of electron and hole-drift regions as in GaAs since the saturated velocity of holes is quite low (0.2×10^7 cm/s in 6H-SiC and 1×10^7 cm/s in the 3C polytype; Trew et al., 1991). Such diodes have been modeled by Trew et al. (1991), who predict the possibility of SiC IMPATTs in the range of 20 to 30 GHz giving about 4 W RF power with 15-20 percent DC-to-RF conversion efficiency.

According to the simulation, the efficiency of SiC devices would be poor compared to IMPATTs of GaAs above 40 GHz (Figure C-10). Low efficiency would offset the advantage of higher thermal conductivity of SiC. The low predicted efficiency for double-drift SiC IMPATTs seems to result not only from assumed high values of parasitic series resistance but also from the high DC-bias voltage needed to sustain avalanche breakdown. These voltages range from 517 V for 35-GHz diodes to 236 V at 94 GHz. Compared with this, RF voltage swings predicted in the simulations are small (~ 80 V at 45 GHz). Such small terminal voltage swings are associated with low efficiency that is given in the large-signal limit approximately by

$$h = 0.72(V_{rf}/V_{dc}) , \quad (C.8)$$

where h is the DC-to-RF conversion efficiency and V_{dc} is the operating voltage neglecting parasitic series resistance (Masse et al., 1985). From Equation (C.8) one finds h = 14 percent.

In order to predict the performance of SiC IMPATT diodes more accurately, numerical simulations should be extended to larger signal. It should be determined if net negative resistance is still available given that the diode conductance magnitude varies roughly as $|G| \sim 1/V_{rf}$. Predictions of higher efficiency might result if higher

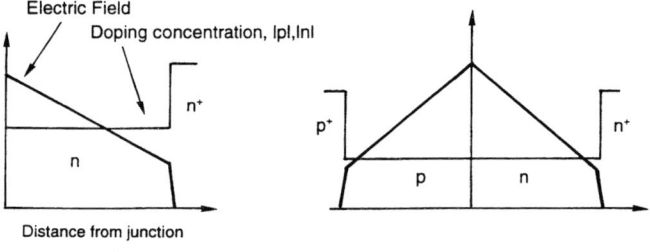

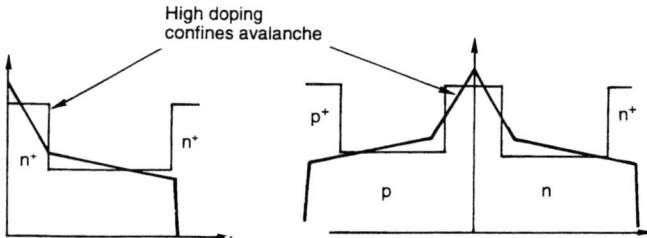

FIGURE C-11 Material structures and electric field profiles possible for IMPATT diodes. Clockwise from upper left: single-drift flat, double-drift flat, single-drift Read, double-drift Read.

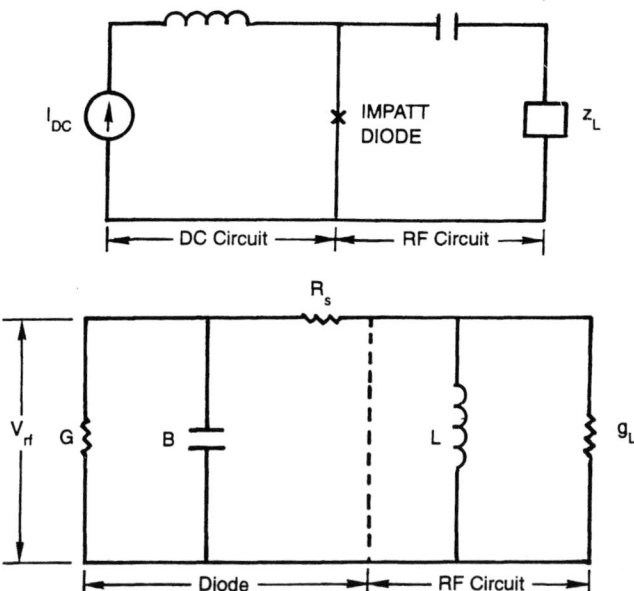

FIGURE C-12 A simplified equivalent circuit for an IMPATT diode embedded in a microwave circuit.

current densities were used in the simulations since resulting values of |G| would be higher.

It is clear that there are some fundamental problems relating to the usefulness of SiC IMPATTs. The devices are complex, and there are many interacting factors to consider. Currently, research is being conducted on SiC IMPATTs. Performance goals at 35 GHz are 12-W continuous-wave with DC-to-RF conversion efficiency of 15 percent. As of this writing, no RF results have yet been published.

Diamond is another wide bandgap semiconductor with potential utility for IMPATT diodes. Diamond has the advantage of electron and hole mobility relatively higher than that of SiC. High operating voltages would be needed at a given frequency because of the low (= 5.5) relative dielectric constant and the high values of avalanche breakdown fields. The high thermal-conductivity of diamond compared with all other wide bandgap materials allows for increased ambient temperature at a given power dissipation at the junction. It was predicted by simulation that diamond IMPATTs have superior RF output power up to 100 GHz when compared with silicon and GaAs (Trew et al., 1991). At 100 GHz, 1.5-W power at 10 percent power-added efficiency is predicted for a double-drift diamond IMPATT. RF performance is predicted to degrade above 100 GHz due to the finite extent of the avalanche region that becomes comparable to the drift-region thickness.

One issue complicating the use of diamond in pulsed operation is the high activation energy of its acceptors (350 mV for boron) and possible donors. There would be a fairly strong temperature dependence of carrier concentration in the contact and drift regions that would vary over the length of a pulse. This effect, coupled with the relatively short thermal time constants for diamond could result in time-varying power output. Another issue relevant to diamond is whether suitable n-type dopant can be found to construct double-drift IMPATTs. Alternatively, p-type layers could be used in single-drift structures. This would not necessarily be a disadvantage for performance since hole-saturated velocity is 10^7 cm/s, a value twice as high as for holes in GaAs. The saturation electric field for holes is around 20 kV/cm in diamond.

A question that relates to all of the devices considered above is whether they offer an advantage over existing devices other than the potential operation at temperatures higher than possible with GaAs or silicon. If modest performance is acceptable at elevated temperatures and the main issue is reliability and the simplification of system heat-sinking requirements, then SiC IMPATTs could be useful. On the other hand, it would be desirable to develop the full potential of diamond IMPATT diodes since they offer performance superior to that from SiC.

EXPECTATIONS FOR WIDE BANDGAP MESFETs

Of all the microwave devices using wide bandgap semiconductors, MESFETs have received the most attention from researchers. MESFETs do not require a high-quality oxide for fabrication, and it is relatively simple to apply micron and submicron gates to the surface of crystals. Ideally, semi-insulating substrates are required, but p-n junctions can be substituted to fabricate devices for testing. Researchers have reported 6H-SiC devices with 0.6-μm gate lengths that had room-temperature DC transconductance of 25 mS/mm and exhibited 14 dB small-signal gain at 1 GHz despite high parasitic source resistance (Palmour et al., 1991, 1993). A power density of 4 W/mm and a total corresponding output power of 65 W is predicted at 10 GHz (Trew et al.,

1991). At the present time, demonstrated performance falls far short of predictions.

Predictions of performance have been based on elaborate numerical models of MESFET operation. To bring the issues into better focus, it is useful to consider the problem in simpler terms using analytical models in which parameters are normalized to materials properties (Shur, 1987). The first objective is to calculate current-voltage characteristics for MESFETs of various materials. These can then be used to estimate power and efficiency from an assumed DC load line and make comparisons. It is more difficult to calculate gain from these simple models, and for this it is necessary to rely on more elaborate time-dependent modeling. Alternatively, gain can be estimated from equivalent circuit models that take into account parasitic resistance and reactance. Expected transistor-operating temperature rise can then be considered in conjunction with gain predictions.

In addition to affecting gain, parasitic resistance can have a profound effect on the current-voltage curves, so this section first considers the intrinsic MESFET, in which behavior depends on the basic velocity versus electric field characteristics. The following analysis is concentrated on SiC as a prototype for all wide bandgap semiconductors. Later sections consider the special properties of other materials.

Current-Voltage Curves

Intrinsic MESFETs fall into three possible regimes, depending on whether or not velocity saturates with drain bias in the channel before pinch-off occurs. The key parameter is $a = F_s L/V_{po}$, where F_s is the field at velocity saturation, L is the channel length, and V_{po} is the intrinsic pinch-off voltage given by

$$V_{po} = qN_D X_c^2/(2e) , \quad (C.9)$$

where q is the electronic charge, N_D is the channel ionized donor density, X_c is the thickness of the channel layer, and e is the material dielectric constant. The parameter a is a measure of the relative importance of velocity saturation and channel pinch-off in determining the drain voltage at which the drain current saturates. Velocity versus field curves indicate that the velocity of electrons in SiC saturates at 8×10^4 V/cm, considerably higher than silicon or GaAs. From its value of a, MESFETs of each material can be represented in parameter space with the axis consisting of channel doping and gate length as shown in Figure C-13. The boundary labeled $a = 0.33$ separates a region (to the left of the line) where current saturation with applied drain voltage results from electron-velocity saturation in the channel. Below the line labeled $a = 3$, current saturation results because the channel pinches off at its drain end due to voltage drop along the channel. Simple analytical models for the MESFET are available for these two cases, while the region between the boundary lines requires a more detailed treatment. Assuming 0.5-μm gates are used, Figure C-13 shows that silicon and GaAs can both be well represented by the assumption of velocity saturation. This is also the case for SiC if the channel doping is held above 3×10^{17} cm^{-3}. A N_d value commonly used for analysis is 2.5×10^{17} cm^{-3}, and it would be a reasonable approximation to assume the velocity-saturation model for SiC. In contrast, GaN requires a higher electric field for electron-velocity saturation (150 kV/cm) so that shorter gates or higher channel doping is required to saturate electron velocity.

In the cases of GaAs, silicon, and SiC, channel saturation can be assumed and the equations describing the current-voltage curves for the MESFET can be written in dimensionless form (Sze, 1969) with all applied voltages normalized to V_{po} and the drain current normalized to $g_o V_{po}$ where

$$g_o = q\mu N_D X_c W/L . \quad (C.10)$$

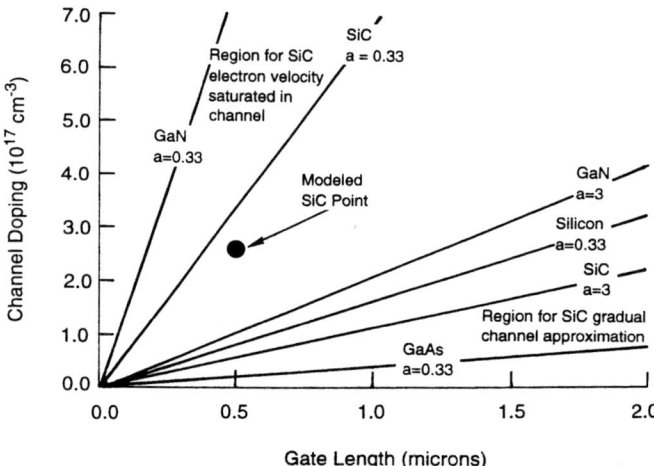

FIGURE C-13 'a' contours for MESFETs of silicon, GaAs, silicon carbide, and gallium nitride.

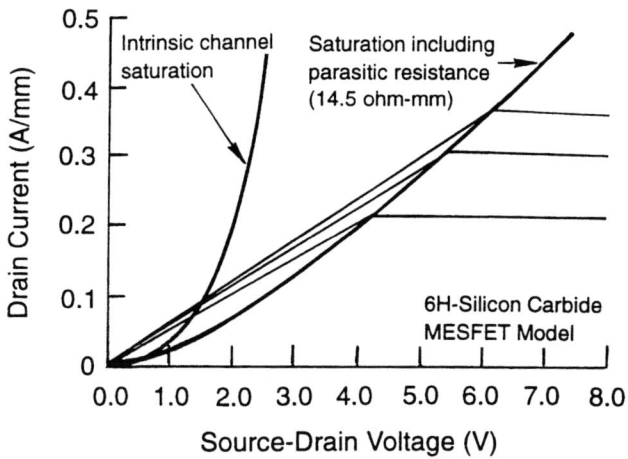
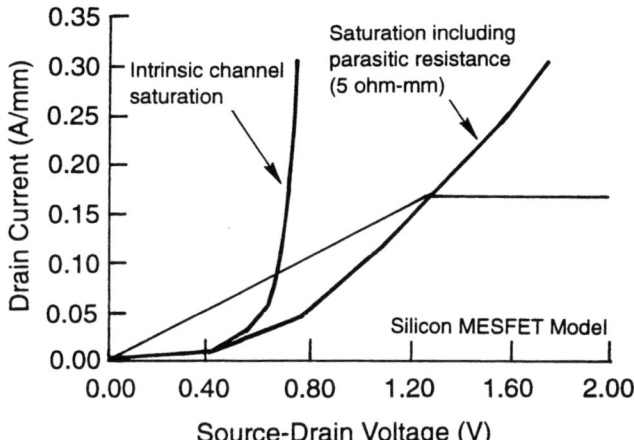
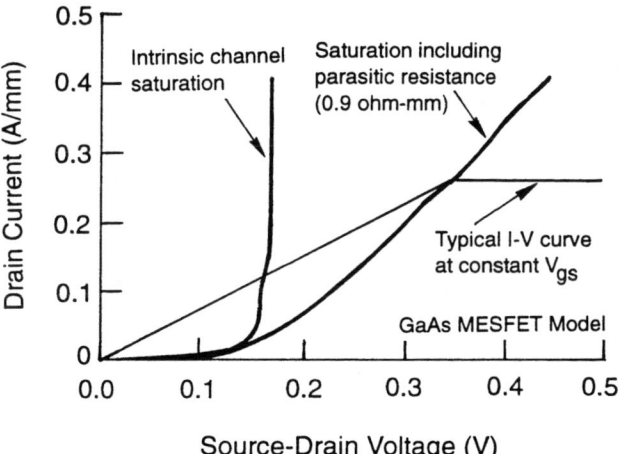

FIGURE C-14 Calculated locus of drain-current saturation for (a) silicon carbide, (b) silicon, and (c) GaAs (with and without parasitic series resistance).

In Equation (C.10), L is the gate length (typically 0.5 μm) and W is the gate width (typically 1 mm). The transconductance of the intrinsic MESFET is also independent of the details of the material but is normalized to g_o. The gate bias, u_g, is shifted by the built-in voltage and normalized to the pinch-off voltage:

$$u_g = (V_{bi} - V_g)/V_{po} , \qquad (C.11)$$

where V_{bi} is the built-in Schottky barrier voltage and V_g is the applied gate voltage. According to Equation (C.11), large built-in voltages typical of wide bandgap semiconductors offset the effect of applied gate voltage by partially depleting the channel at zero applied-gate bias.

Table C-2 summarizes the results of the model in which parameters are calculated for typical design values for three materials and a 0.5-μm gate length. Included in the lower portion of the table are assumed and calculated parasitic resistances based on an N^+ contact doping density of 5×10^{18} cm^{-3}. Chosen dimensions for the electrode spacings are shown in the table. These dimensions would be typical of test devices made with standard processing technology.

From the modeled values in Table C-2, the saturated drain current and associated drain voltage can be calculated. Modeled saturation curves are shown for the three semiconductors in Figure C-14. For each semiconductor, two current saturation curves are shown: (1) the intrinsic saturation in the absence of parasitic resistance and (2) the extrinsic saturation using typical values of parasitic resistance. Points on these curves can be reached by varying the gate bias on the MESFET and increasing the drain bias until current saturates. A few familiar typical current-voltage curves at constant gate bias are shown for reference.

A noteworthy feature of the I-V curves is that I_{dss}, the maximum channel current at $V_g = 0$, is similar for SiC and the other materials even though the saturation velocity of electrons in SiC is greater. This is so because the built-in voltage from the gate contact is higher for the large bandgap material. The precise value of this voltage depends on the Schottky metal used for the gate, but it is assumed here that the voltage is equal to about one-half the bandgap voltage in each case. In operation, higher peak currents can be obtained for all materials by driving the gate somewhat into forward bias. This is not recommended since doing so often has reliability implications. Another option for increasing the maximum current in SiC would be to use heavier doping levels. This would reduce breakdown voltage but would have the added beneficial effect of reducing the parasitic resistance

TABLE C-2 Assumed and Calculated MESFET Current-Voltage Model Parameters

	SiC	Silicon	GaAs
Channel doping (cm^{-3}), N_d	2.5×10^{17}	2.5×10^{17}	2.5×10^{17}
Gate width (mm), W	1	1	1
Active layer thickness (μm), X_c	0.2	0.2	0.2
Gate length (μm), L	0.5	0.5	0.5
Mobility (cm^2/V·s), u	250	600	4,000
Relative dielectric constant, e	10.0	11.8	12.8
Gate built-in voltage (V), V_{bi}	1.63	0.50	0.70
Intrinsic pinch-off voltage (V), V_{po}	9.0	7.9	7.1
Intrinsic transconductance (S), g_o	0.4	1.0	6.4
Electron saturated velocity (cm/s), v_s	2.0×10^7	1.0×10^7	0.6×10^7
Velocity saturation field (V/cm), F_s	8.0×10^4	1.7×10^4	0.35×10^4
Channel saturation parameter, α	0.45	0.11	0.02
Metal contact resistance ($\Omega \cdot$cm^2) R_{co}	10.0×10^{-6}	1.0×10^{-6}	1.0×10^{-6}
N$^+$ layer sheet resistance ($\Omega/\square$), R_{sq}	843	100	25
Transfer length (μm), L_t	1.1	1.0	2.0
Net contact resistance (Ω), R_o 1 μm contact length	1.3	0.13	0.11
Source-gate access resistance (Ω), R_{gs} 0.5 μm gate-source spacing	2.5	1.0	0.16
Gate-drain access resistance (Ω), R_{gd} 1 μm gate-drain	5.00	2.08	0.31

SOURCE: Morkoc et al. (1994).

between the gate to the source and drain electrodes respectively. The limits of doing this for SiC should be explored in experimental studies.

A comparison of the curves in Figure C-14 shows that the saturation voltages for SiC are much higher than for GaAs or silicon and that at high currents this is dominated by the effects of parasitic resistance. Even if parasitic resistance could be reduced to zero, drain current at saturation would still be an order of magnitude greater for SiC than for GaAs. This is due to the higher electron saturation fields for SiC.

When the saturation voltages are high, higher DC bias voltages must be used to obtain efficient RF amplification. Fortunately, such voltages can be achieved in wide bandgap semiconductors. For the purpose of further modeling of this effect, Figure C-14 defines saturation resistance, R_s, as an effective lower bound for the drain voltage during high-frequency operation when the peak current is near saturation.

Because of its importance in determining the current-voltage characteristics for the MESFET, it is worthwhile to consider the origin of the parasitic resistance in SiC in order to identify the technological problems to be solved.

Figure C-15 shows a simple model for ohmic contact and channel resistance contributions to MESFET source resistance. The specific contact resistivity is represented by resistors in parallel while the resistance of the material underneath the contact conducts increasing current density

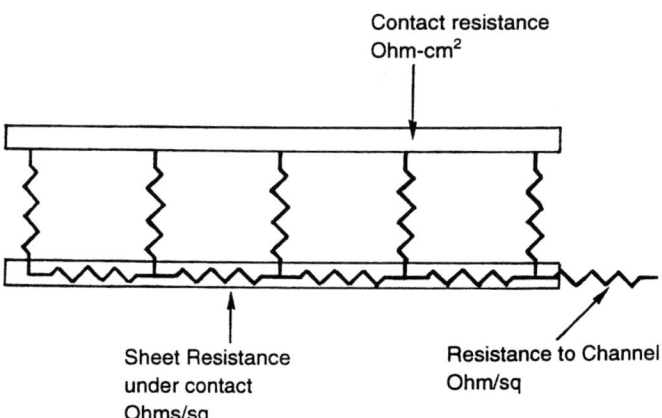

FIGURE C-15 A simple model for ohmic contact and channel resistance contributions to MESFET source resistance.

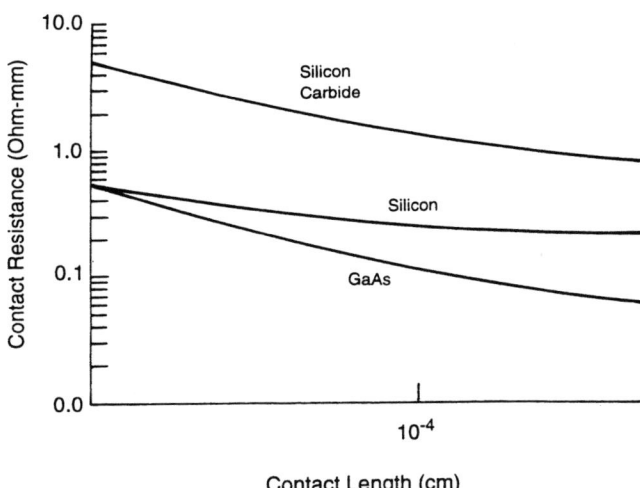

FIGURE C-16 Contact resistance calculated as a function of contact length for three materials. SOURCE: Palmour et al. (1993).

toward the MESFET channel. It can be shown from this model that the contribution of the contact to the resistance in ohms is given by

$$R_o = (Z/W) \coth(L/L_t) , \qquad (C.12)$$

where $Z = (r_c R_{sq})^{1/2}$ and $L_t = (r_c/R_{sq})^{1/2}$, with r_c being the specific contact resistivity (in units of $\Omega \cdot cm^2$ while W is the contact width in centimeters) and R_{sq} being the sheet resistance of the contact layer in Ω/square. For a contact of length $L \gg L_t$, further increase in L does not result in a decrease in contact resistance. In this limit, contact resistance is approximately given by $R_o = Z/W$. This is illustrated in Figure C-16 where contact resistance is calculated as a function of contact length for three materials, assuming published values of mobility (Rahman and Furukawa, 1992; Palmour et al., 1993) and n-type doping of 5×10^{18} cm^{-3}. In Figure C-17, contours of Z and L_t are plotted on the r_c-R_{sq} plane. Typical points for GaAs, silicon, and SiC are included in the figure. For particular transistors where $L \gg L_t$, the product $R_o W$ can be read directly as Z in the figure. It can be seen that both GaAs and SiC benefit by making the contacts larger than 1 μm long. Note that in technology development, it is proper to concentrate efforts on both reducing contact resistivity as measured in $\Omega \cdot cm^2$ as well as reducing semiconductor layer resistivity.

If contacts are to survive elevated temperatures, refractory metals must be used. Contact resistivities at elevated temperatures are given for refractory metals in Table C-3 (Shur et al., 1993). These values are considerably higher than that assumed for the 25 °C analysis. There will be a corresponding increase of the saturation voltage in the current-voltage curves. Also contributing to the saturation voltage at elevated temperatures is the fact that at doping concentrations typical of MESFET channels (2 to 5×10^{17}), mobility will decrease as temperature increases beyond 25 °C (Goetz et al., 1993; Shur et al., 1993). The increase is due to increasing phonon scattering. One difference between the wide bandgap semiconductors compared with GaAs and silicon is the larger optical phonon energy in the wide bandgap material. This difference implies a less rapid drop-off in mobility in the wide bandgap case, which might be viewed as an advantage. The increase in saturation voltage with increasing temperature will effect the power performance of the MESFET but, as discussed later in this section, will have a much greater effect on the gain and efficiency of the transistor.

Power and Efficiency

From the predicted current-voltage relationships for various materials, power performance of MESFETs can be compared. Amplifier operation is described by the switching of current from the drain supply alternately through the MESFET and the external load. The voltage at the gate supplied by the power source modulates the saturated current of the channel. In this process, the point on the current-voltage plane representing the instantaneous state at the MESFET drain moves along one of the trajectories (loadlines) shown in Figure C-18. Class A operation, for which highest power is obtained, is illustrated. The loadlines in the figure are appropriate to conditions of maximum RF input power level such that the gate is not driven into forward conduction or reverse breakdown at any point in the RF cycle. The output power is then area-bounded by the I-V axis and the trajectory. In some computer numerical simulations, overdrive of the gate is allowed along with nonlinearities and resulting harmonics resulting in prediction of higher power densities. However, for simplicity and emphasis of the main points, this possibility is disallowed in the present analysis.

A set of parameters can be defined that can be used with the current-voltage curves to model expected power and drain efficiency of the transistor from knowledge of the DC bias voltage and the saturation resistance, R_s (Adlerstein and Zaitlin, 1991). The model variables for power output are independent of material when they are normalized to parameters characteristic of the bias point and the current-voltage curves. External load resistance ($r_1 = 1/g_1$) is normalized to R_s, drain current is normalized to $I_s = V_{DC}/2R_s$, and power is normalized to $P_s = V_{DC}^2/(2R_s)$. Calculated values for these parameters obtained from the power model are shown in Table C-4. MESFET size is taken to be 480 μm, similar to that used in large numerical simulation (R.J. Trew, personal communication, 1994). It is seen from the characteristic power P_s that some of the advantage of higher operating voltage in SiC is mitigated by the accompanying higher value of R_s. Nevertheless, V_{DC} can still be made high enough to give a power advantage to SiC.

In Table C-4, an upper bound on the operating voltage is used, obtained from the breakdown electric fields for SiC, silicon, and GaAs using the equation

$$V_b = (E_m^2 e/2qN_D) , \qquad (C.13)$$

where E_m is the critical electric field at junction. For GaAs, the gate to drain breakdown, V_b, is often higher due to Gunn domain formation near the drain end of the gate. For comparison in Table C-3, however, the operating voltage is $(V_b - R_s I_{dss})/2$, which is traditionally chosen to maximize the power and efficiency of MESFETs. Likewise, $I_{dc} = I_{dss}/2$ was chosen to maximize the area bounded by the I-V trajectory. SiC MESFETs of two designs are listed in Table C-4. The case in the second column is taken to have channel doping of 2×10^{17} as assumed for the MESFETs of the other materials. For this 480-μm-gate-width MESFET, output is around 3.9 W, ten times higher than the power available from the comparable GaAs MESFET in the table.

To realize the high-power level predicted for this SiC design, high RF voltage amplitude is needed, and this implies that high output impedance is required. The passive microwave circuits needed to achieve these impedances could be narrow band or loss (Vandelin, 1982). Furthermore, available substrate material may be leaky and result in a relatively high output conductance for the transistor. Output conductance is further increased by gradual saturation of electron velocity in the channel. A high output conductance would imply that a low value of load resistance would be required for a conjugate match condition giving maximum gain. This is a condition in conflict with maximizing power. Making the transistor larger to reduce the power-load impedance would not diminish the ratio between the gain match and power match since both would scale with device size. Gain would already be diminished by the high parasitic source resistance in a SiC MESFET through which gate charge must be provided by the microwave exciter during the RF cycle.

The discrepancy between power match at high operating voltage and gain match for SiC appears to be common to all the wide bandgap materials due to their characteristically high parasitic series resistance and high saturation voltages. Compromises must be found to optimize the load. The situation is illustrated schematically in Figure C-18. At low bias voltage, loadline "A" gives maximum gain while loadline "B" gives maximum power. At a higher bias-voltage loadline, "C" maximizes both gain and power (assuming device output conductance is not different at this bias). At the highest applied voltages, the load resistance for maximum power can become too high and there will be excessive mismatch at the output resulting in reduced gain. The discrepancy in required load impedances for power and gain could account for the less-than-optimum results reported for power tests in discrete devices of wide bandgap material; tuning, operating voltage, and operating current must all be compromised in the test.

One way to decrease the power-load resistance at a given bias voltage is to increase I_{dss} and reduce the

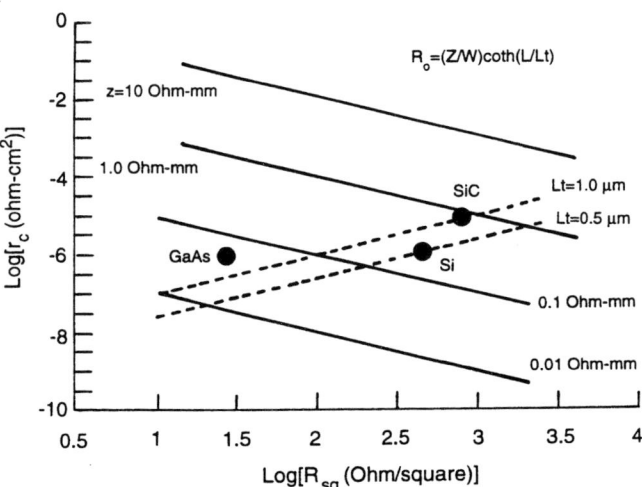

FIGURE C-17 Contours of constant Z plotted on r_c-R_{sq} plane. SOURCE: Rahman and Furukawa (1992), © 1992 IEEE.

Appendix C: High-Temperature Microwave Devices

TABLE C-3 Listing of Several Refractory Metallizations on SiC and their Contact Resistivities

Material	Type	Metal	Doping Density (cm^{-3})	Contact Resistivity ($\Omega \cdot cm^2$)	Maximum Useful Temperature (°C)
3C-SiC	n	Ti/TiN/Pt/Au	10^{16}-10^{17}	1.2×10^{-4}	650
3C-SiC	n	W/Pt/Au	10^{16}-10^{17}	1.5×10^{-4}	650
6H-SiC	n	TiN	1.6×10^{18}	4×10^{-2}	550
6H-SiC	p	3C-SiC/Al/Ti	1-3 $\times 10^{18}$	2×10^{-5}	<450
6H-SiC	p	Al/Ti	2×10^{19}	1.5×10^{-5}	<450

SOURCE: Shur et al. (1993).

operating voltage of the transistor. This will result if higher channel doping is used as shown in the first column of Table C-4 for the case where N_d is increased by a factor of two to 5×10^{17} cm^{-3}. This change increases I_{dss} by about a factor of two. The DC bias voltage is reduced by a factor of two as necessitated by the lower gate drain breakdown voltage at the higher doping level. The design change decreases the power load by a factor of four, bringing it into the range needed for GaAs devices of comparable size. Power density is now around 4.3 W/mm. As expected, the change in doping reduces the drain efficiency compared with the former SiC case from 46 percent to around 37 percent. This is due to the high saturation resistance of SiC despite a decrease in the channel access resistance since there is still a high knee voltage at the increased I_{dss}. The power added efficiency will decrease only slightly since the gain (taken to be 8 dB and 10 dB respectively) at full power should be higher due to the improved match achieved with the power load (Trew et al., 1991). In the detailed design of wide bandgap MESFETs, tradeoffs between the parameters discussed above should be taken into account to optimize performance.

MESFET Gain

Up to this point, this appendix has considered the relationship between output power and the current-voltage characteristics of SiC MESFETs compared with silicon and GaAs. Parasitic series resistance was found to be detrimental to drain efficiency, but high power density could still be achieved given the possibility of operating at high bias voltages. To make SiC MESFETs useful, high gain in addition to high power must be achieved.

Gain, defined as the ratio between RF input and RF output power, is a decreasing function of frequency. The frequency response of the basic MOSFET under small-signal conditions is represented by the equivalent circuit shown in Figure C-19 (Englemann and Liechti, 1977). To a good approximation, at X-band and below, the input resistance, R_i, is given by

$$R_i = R_i^o + R_g + R_s , \qquad (C.14)$$

where R_i^o is the intrinsic channel resistance, R_g is the parasitic gate resistance, and R_s is the parasitic source resistance. The drain resistance, R_d, is explicitly shown to conform with the reference plane for the DC current–voltage curves.

The frequency for unity current gain, f_t is given by

$$f_t = g_m/(2\pi C_{gs}) = 1/(2\pi t_g) , \qquad (C.15)$$

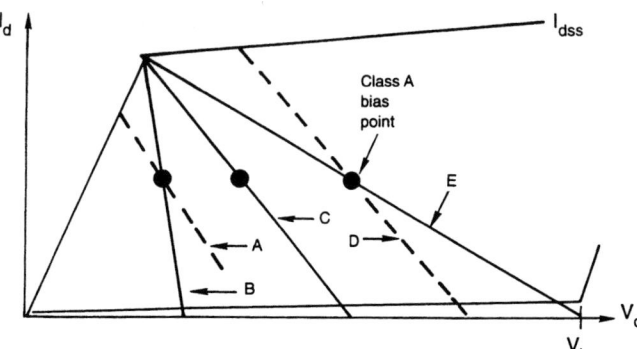

FIGURE C-18 Representation of current-voltage curves for a MESFET and typical loadlines for Class A operation.

TABLE C-4 Assumed and Calculated MESFET Power Model Parameters

	SiC	SiC	Silicon	GaAs
DC bias voltage (V), V_{dc}	32	75	4	8
Saturation current density (A/mm), I_{dsso}	0.96 ($N_d = 5.0 \times 10^{17}$)	0.48 ($N_d = 2.5 \times 10^{17}$)	0.3	0.41
Specific saturation resistance ($\Omega \cdot$mm), R_{so}	9	14.5	5	0.9
MESFET gate width (mm), Z	0.48	0.48	0.48	0.48
Gated saturation current (A), I_{dss}	0.461	0.230	0.144	0.197
Saturation resistance (Ω), R_s	18.8	30.2	10.4	1.9
Load conductance (S), g_l	0.010	0.002	0.029	0.013
Load resistance (Ω), r_l	101	590	35	78
Model parameter, $g_l R_s$	0.185	0.051	0.300	0.024
Drain efficiency, E_{fd}	0.365	0.454	0.313	0.477
Model parameter (W), P_s	27.3	93.1	0.77	17.1
RF power delivered to load (W), P_{dis}	2.7	3.9	0.1	0.4
Drain current (A), I_{dc}	0.231	0.115	0.072	0.098
Dissipated power (W), P_l	4.68	4.72	0.20	0.41
Thermal conductivity (W/cm$\cdot$°C), C_{th}	5	5	1.5	0.5
Channel temperature Rise (°C), DT	63.5	63.9	9	55.5
Gain (dB), G	10	8	10	10
Power added efficiency (%), PAE	0.329	0.382	0.281	0.429

SOURCE: Shur et al. (1993).

where C_{gs} is the gate-to-source capacitance, g_m is the MESFET transconductance, and t_g is the transit time under the gate. Potentially, f_t is higher in SiC where the saturated velocity is greater than GaAs, particularly at drain voltages just above the saturation value. At the high bias voltages required for efficient and high-power microwave operation, SiC frequency response is expected to decrease. This results since the effective gate length should include a distance beyond the gate where the channel is depleted due to the applied drain voltage. Despite the potential of SiC, measured values of f_t so far have been much lower than expected. The reason for the discrepancy is not known. It might be speculated that the presence of traps in the material, particularly between the gate and drain contacts, might significantly increase the charging time of the gate. Alternatively, the surface between the contacts may charge through high resistance and not be responsive to high-frequency voltages. Such effects would not be evident in the DC I-V curves but could be revealed by pulsed I-V measurements as they have been for GaAs MESFETs (Platzker et al., 1990).

Given a value of f_t, one can utilize as a gain figure of merit the unilateral gain $U = (f_{max}/f)^2$. In terms of the equivalent circuit of Figure C-19, f_{max} is given by

$$f_{max} = f_t / \{2[(R_i G_{ds}) + 2\pi f_t R_g C_{dg}]^{1/2}\} . \quad (C.16)$$

Equation (C.16) shows how parasitic series resistance reduces gain.

Compared with GaAs, G_{ds} (the output conductance) is larger in SiC where substrate leakage can be greater and the rate of velocity saturation is lower. Furthermore, the

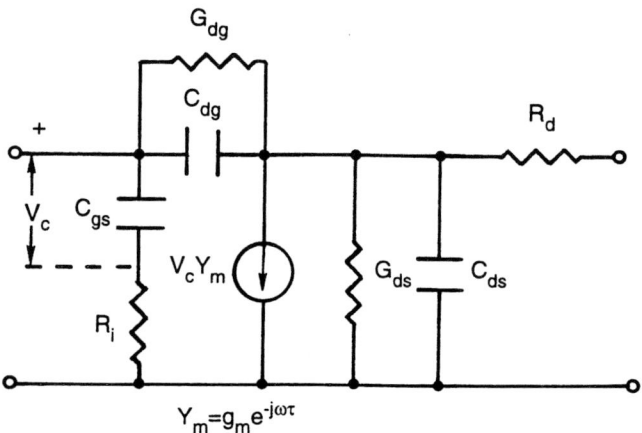

FIGURE C-19 Small signal equivalent circuit for a MESFET.
SOURCE: Englemann and Liechti (1977), © 1977 IEEE.

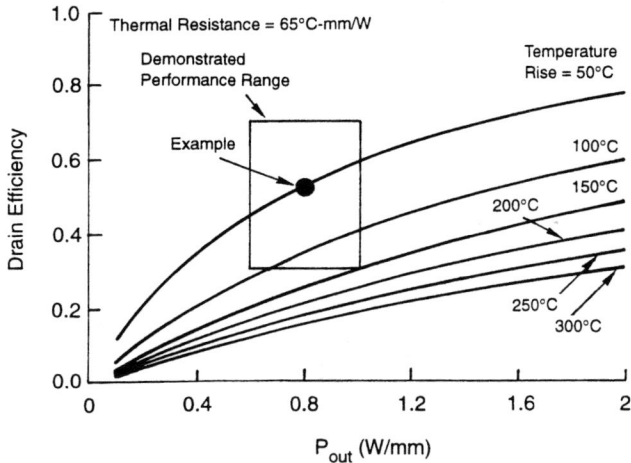

FIGURE C-20 Contours of constant temperature rise in the GaAs MESFET channel. SOURCE: Wemple and Huang (1982).

parasitic source resistance in SiC is considerably larger than for GaAs with comparable dimensions. Thus, if SiC is to exhibit gain comparable to or higher than GaAs, f_t would have to be much higher. This might imply limitations on the drain voltage and resulting limitations in power. The situation regarding gain is quite complicated, involving both fundamental material properties and technology issues. It is concluded that the ultimate frequency response of SiC MESFETs must be established by experiment.

Thermal Properties of SiC MESFETs

SiC has a thermal conductivity approximately ten times that of GaAs. This section attempts to explain, based on simple models, whether or not the higher thermal conductivity of SiC offers a performance advantage. The section also shows plots of the available experimental data relative to these predictions. To determine the performance advantage, it is noted that the temperature rise of MESFETs depends on the power dissipation and the thermal resistance of the transistor. The temperature rise in the channel can be expressed in terms of the output power, gain, and power-added efficiency according to $\Delta T = q_{th} P_{dis}$, where P_{dis} is given by Equation (C.2). Contours of constant temperature rise in the channel are plotted in Figure C-20 for GaAs and in Figure C-21 for SiC. For comparison, gain is taken to be 10 dB for which the dissipated power due to the RF input is negligible. The thermal resistance normalized to total gate width for GaAs MESFETs is estimated to be around 65 °C/W·mm,

while that for SiC is reduced in proportion to the thermal conductivity of the material and is taken to be 6.5 °C/W·mm (Wemple and Huang, 1982). Consider first the GaAs contours. As power density increases, drain efficiency must increase to maintain constant operating temperature rise. The example cited in Table C-4 predicts the power point in the figure for Class A operation. The range of experimentally demonstrated results from a literature survey is seen to bracket the example in both power and efficiency (Huang, 1993). Typical transistors show a temperature rise of 50 °C or less, indicating that GaAs MESFETs are electronically limited by available drain current and voltage rather than thermally limited. Reducing the thermal resistance of GaAs MESFETs may have a slight beneficial effect on the performance of the transistors due to higher mobility, higher electron-saturated velocity, and lower leakage currents at lower temperatures, but it is not expected to be a major benefit.

For SiC, the constant temperature rise contours are plotted in Figure C-21. Note that for a given temperature rise, higher power densities and lower drain efficiencies are allowed. Plotted on this graph are the calculated power and drain efficiencies from column 1 of Table C-3, the prediction of the model of Trew et al. (1991) for 6H-SiC, and the experimental result of Westinghouse (Clarke et al., 1993). In all cases, the temperature rise is very modest—less than 80 °C. The committee's conclusion is that as for GaAs, the SiC MESFETs are electronically limited but benefit from the higher thermal conductivity of the material in that the transistors can be pushed to higher power density even with relatively low efficiency without thermal consequence.

Wide Bandgap MESFETs at Elevated Temperatures

Silicon Carbide

Although the foregoing discussion focuses on comparison of GaAs MESFETs with SiC, the major conclusions would apply to a number of other wide bandgap materials, such as GaN and diamond, where low mobility and high contact resistance exist to some degree. In the analysis above, the values for these parameters were taken, optimistically, as those observed at 25 °C. As the channel temperature increases, mobility decreases. This is illustrated in Chapter 3 in Figure 3-1 as calculated for electrons in undoped 6H-SiC and 3C-SiC (Shur et al., 1993). In going from 25 °C to 200 °C, the mobility of 6H-SiC decreases from 420 cm^2/V·s to around 120 cm^2/V·s. The room-temperature mobility is expected to be lower (~250 cm^2/V·s) for SiC doped in the 10^{17} cm^{-3} range where impurity scattering is important. This is particularly true for compensated or poorly activated material. Above room temperature, phonon scattering dominates and the value of mobility at 200 °C is expected to be about 100 cm^2/V·s more or less, independent of doping (Goetz et al., 1993).

In the foregoing MESFET analysis, the lowest reported values of contact resistance is used. At elevated temperatures, refractory metals would be required. Those demonstrated so far, which would be useful at the highest temperatures, would be expected to have resistivities well above 10^{-5} W·cm^2. Continuing efforts are required to lower the contact resistance of refractory metals.

Gallium Nitride

Gallium nitride is an alternative candidate for high-temperature MESFETs. GaN MESFETs have been reported in the literature with a 0.6-μm-thick channel layer (Khan et al., 1993). Channel doping was 10^{17} cm^{-3} and with a gate length of 4 μm, a DC transconductance of 23 mS/mm was obtained. This material is promising because it has a higher mobility than SiC. Like GaAs, GaN has a region of negative differential mobility with quite a high peak electron velocity at 2.3 x 10^7 cm/s and a high saturated velocity at around 1.4 x 10^7 cm/s. At 200 °C, electron mobility is twice that of SiC at the same temperature (Figure 3-2 in Chapter 3). The saturated velocity of the material is unchanged, although the peak velocity decreases steadily (Shur et al., 1993). The committee concludes that the projected gain and efficiency of GaN MESFETs will be slightly higher then their SiC counterparts at elevated temperatures once refractory metals are used.

In an analogy with HEMT design, one option suggested for GaN is the use of alloy and heterojunction material to improve the performance of FETs at elevated temperatures. SiC can be used for substrates upon which to grow hetero-epitaxially GaN or alloys containing aluminum or indium. The thermal conductivity of SiC is about 3.8 times as large as GaN so that MESFETs made from hetero-epitaxial material could have higher power dissipation density. Another option is to fabricate "high electron-mobility" transistors by creating a two-dimensional electron gas at the interface between GaN and AlGaN (Figure C-22). The doping level of the conducting channel can be reduced to a minimum. This could improve mobility in the channel, provided temperatures were not so high that phonon scattering dominated. A heterojunction FET similar in size to the MESFET described above was fabricated with a GaN channel and an AlGaN cap layer (Khan et al., 1992). A DC transconductance of 28 mS/mm was obtained. Further experimental work would be required to assess the potential of heterojunction MESFETs in nitride systems. The usefulness of AlN in homojunction or heterojunction transistors should be determined by continuing experimentation.

Diamond

This section considers prospects for diamond MESFETs. Diamond has many favorable properties. At 25 °C, electron and hole mobility, at least in undoped single crystals, are considerably higher than for the other wide bandgap materials. The material has a high breakdown electric field and a high associated avalanche breakdown voltage. Diamond has a low dielectric constant that minimizes parasitic capacitance associated with electrodes and depleted layers. It has a saturated velocity for electrons of around 2.7 x 10^7 cm/s from which potentially high values of f_t should be obtained. For example, $f_t \sim$ 100 GHz should be possible with breakdown voltage of several hundred volts (Geis et al., 1987). If MESFETs

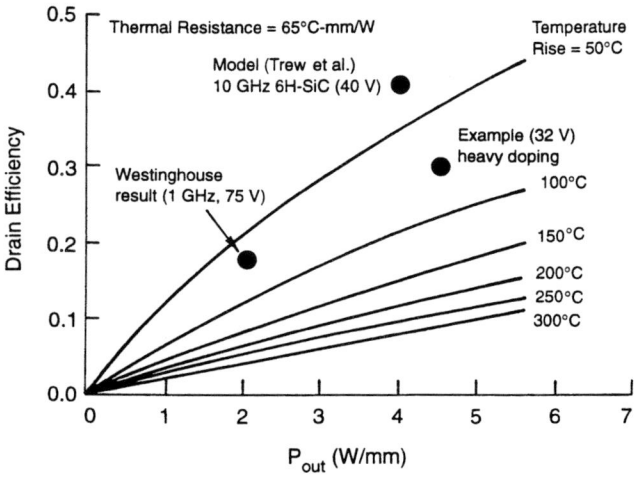

FIGURE C-21 Contours of constant-temperature rise in the SiC MESFET channel. SOURCE: Wemple and Huang (1982).

Appendix C: High-Temperature Microwave Devices

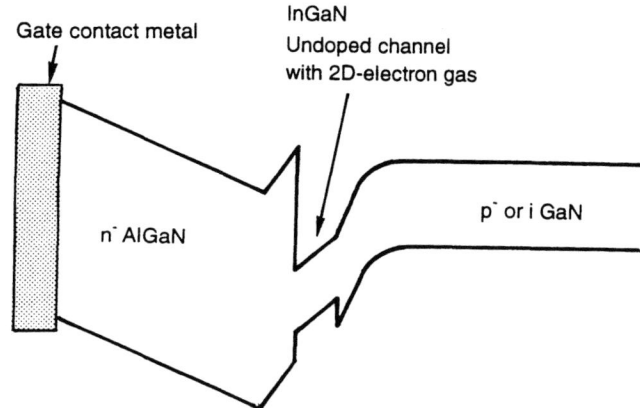

FIGURE C-22 A MODFET transistor with a two-dimensional electron gas at the interface between GaN and AlGaN. SOURCE: Khan et al. (1992).

can be fabricated in n-type diamonds, high-frequency performance better than that predicted for SiC could be obtained. Extensive computer numerical simulation of diamond MESFETs was conducted by Trew et al. (1991). They considered n-type transistors and found that diamond could produce a power density of 6 W/mm compared with 1 W/mm for a similar GaAs MESFET. Power-added efficiencies were comparable in the 40-50 percent range, with diamond being 10 percent more efficient than the GaAs device due primarily to a higher operating voltage. They also predicted that gain for the n-type diamond transistor would be 2 to 3 dB higher than for GaAs. This is presumably due to the lower gate-to-drain feedback capacitance resulting from the lower dielectric constant of diamond. This gain could be traded off against device size allowing for larger MESFETs constructed of diamond. Note that the option of increasing the device size significantly compared to GaAs is unique to n-type diamond since it has a much lower dielectric constant and higher carrier mobility than other wide bandgap semiconductors.

It has proven difficult to dope diamond n-type. Accordingly, MESFETs fabricated to date have had p-type channels. Such devices have given relatively low transconductance of 2 μS/mm at room temperature, increasing to 0.67 mS/mm at 400 °C (Geis et al., 1987). There was an accompanying increase in maximum channel current at the elevated temperatures that is attributed to increasing activation of acceptors (boron) that have activation energy of around 350 meV.

In determining material conductivity as a function of temperature, the increased number of holes overcomes the decrease in mobility, which is a strongly decreasing function of temperature. Shin et al. (1993) have established a model using a harmonic balance technique and a numerical simulator to compare the expected performance of p-type diamond MESFETs with that of n-type (nitrogen-doped) SiC at elevated temperatures. They assumed a mobility dependence on temperature of T^{-g} where $g = 1.3$ for 6H-SiC and $g = 2.8$ for boron-doped diamond. In both cases, contact resistivities were assumed to be 1×10^{-5} $\Omega \cdot$cm^2. For the SiC MESFET, RF performance was found to be near optimum at 25 °C with 3.5 W/mm. Performance degraded with increasing temperature. Gain at 8 GHz was predicted to be 16.5 dB and power-added efficiency was 44 percent for a device with 1-mm total gate length. In contrast, increasing the temperature of the diamond MESFET resulted in improved performance up to 680 °C. Maximum channel currents were considerably lower in diamond. At 500 °C, the p-type diamond MESFET was predicted to give power density of 0.75 W/mm at 5 GHz with 33 percent power-added efficiency and around 8-dB gain. It was concluded that SiC MESFETs were preferred over p-type diamond MESFETs at elevated temperature.

One method that has been suggested for increasing the channel current of p-type diamond MESFETs is to increase the doping density. This can be done near room temperature, however, only with sacrifice of mobility due to severe impurity scattering of holes. For example, at 25 °C, an atomic boron concentration of 10^{19} cm^{-3} is required to obtain an activated carrier concentration of 10^{15} cm^{-3}. The resulting mobility at this concentration is reduced by a factor of eight from that obtained for undoped material.

Despite the difficulties described above in using diamonds for MESFETs, the material still has considerable promise. Future work should be aimed at finding alternative doping methods, particularly for n-type material, and developing contacts that withstand high temperatures and have low resistivities.

REFERENCES

Adlerstein, M.G., and E. Moore. 1981. Microwave properties of GaAs IMPATT diodes at 33 GHz. Pp. 375-384 in Proceedings of the 8th Biennial Conference on Active Microwave Semiconductors and Circuits. Ithaca, New York: Cornell University Press.

Adlerstein, M.G., and M. Zaitlin. 1991. Cut-off operation of heterojunction bipolar transistors. Microwave Journal 34(9):114-125.

Clarke, R.C., R.H. Hopkins, C.D. Brandt, M.C. Driver, D.L. Barrett, A.A. Burk, G.W. Eldridge, H.M. Hobgood, J.P. McHugh, P.G. McMullin, R.R. Siergiej, and S. Sriram. 1993. Paper presented at the 1993 IEEE/Cornell Conference on Advanced Concepts in High Speed Semiconductor Devices and Circuits, Ithaca, New York.

Englemann, R., and C. Liechti. 1977. Bias dependence of GaAs and InP MESFET parameters. IEEE Transactions on Electron Devices ED-24(11):1288-1296.

Gao, G.B., J. Sterner, and H. Morkoc. 1994. High-frequency performance of SiC heterojunction bipolar transistors. IEEE Transactions on Electron Devices 41(7):1092.

Geis, M.W., D.D. Rathman, D.J. Ehrlich, R.A. Murphy, and W.T. Lindley. 1987. High-temperature point-contact transistors and Schottky diodes formed on synthetic boron-doped diamond. IEEE Electronic Devices Letters 8(8):341-343.

Goetz, W., A. Schoner, G. Pensl, W. Suttrop, W.J. Choyke, R. Stein, and S. Leibenzeder. 1993. Nitrogen donors in 4H-silicon carbide. Journal of Applied Physics 73(7):3332-3338.

Hobgood, H.M. 1993. Growth of Large Diameter SiC Crystals. Presentation to the Committee for High-Temperature Semiconductor Devices, Washington, D.C., September 29-30.

Huang, J. 1993. Proceedings of the MTT-S Microwave HBT and HEMT Workshop, Atlanta, Georgia.

Johnson, A. 1965. Physical limitations on frequency and power parameters of transistors. RCA Review 26:163-177.

Kelner, G., S. Binari, K. Sleger, and H. Kong. 1987. Beta-SiC MESFETs and buried-gate JFETs. IEEE Electron Devices Letters 8(9):428-430.

Kelner, G., M. Shur, S. Binari, K. Sleger, and H.S. Kong. 1989. High transconductance SiC buried-gate JFETs. IEEE Transactions on Electron Devices 36:1045-1049.

Kelner, G., S. Binari, M. Shur, and J. Palmour. 1991. High temperature operation of alpha-silicon carbide buried-gate junction field-effect transistors. Electronics Letters 27:1038-1040.

Keyes, R.W. 1972. Figure of merit for semiconductors for high speed switches. Proceedings of the IEEE 60:225.

Khan, M.A., R.A. Skogman, J.M. Van Hove, D.T. Olson, and J.N. Kuznia. 1992. Atomic layer epitaxy of GaN over sapphire using switched metalorganic chemical vapor deposition. Applied Physics Letters. 60:3027.

Khan, M.A., J.N. Kuznia, A.R. Bhattarai, and D.T. Olson. 1993. Metal semiconductor field effect transistor based on single crystal GaN. Applied Physics Letters 62:1786-1787.

Masse, D., M.G. Adlerstein, and L.H. Holway. 1985. Millimeter-wave GaAs IMPATT diodes. Pp. 291-370 in Infrared and Millimeter Waves, Vol. 14 K.J. Button, ed. New York: Academic Press.

Morkoc, H., S. Strite, G.B. Gao, M.E. Lin, B. Sverdlov, and M. Burns. 1994. A review of the large bandgap SiC, III-V nitride, and ZnSe based II-VI semiconductor device technologies. Journal of Applied Physics Review 76(3):1363-1398.

Muench, W., P. Hoeck, and E. Pettenpaul. 1977. Silicon carbide field effect and bipolar transistors. Pp. 337-339 in IEEE Digest, International Electron Device Meeting.

Palmour, J.W. 1993. Design and Fabrication of SiC Devices. Presentation to the Committee on Materials for High Temperature Semiconductor Devices, Washington, D.C., September 30.

Palmour, J.W., S. Kong, D.G. Waltz, J.A. Edmond, and C.H. Carter. 1991. 6H-SiC transistors for high temperature operation. Transactions of the First International High Temperature Electronics Conference. pp. 511-518.

Palmour, J.W., J.A. Edmond, H.S. Kong, and C.H. Carter, Jr. 1993. 6H-silicon carbide devices and applications. Physica B 185:461-465.

Platzker, A., A. Palevsky, S. Nash, W. Struble and Y. Tajima. 1990. Characterization of GaAs devices by a versatile pulsed I-V measurement system. Pp. 1137-1142 in IEEE MTT-S International Microwave Symposium Digest, Vol. 3.

Rahman, M.M., and S. Furukawa. 1992. Silicon carbide turns on its power. IEEE Circuits and Devices 22-26.

Shin, M.W., G.L. Bilbro, and R.J. Trew. 1993. High temperature operation of N-type 6H-SiC and P-type diamond MESFETs. Pp. 421-430 in Proceedings of the IEEE/Cornell Conference on Advanced Concepts in High Speed Semiconductor Devices and Circuits.

Shur, M. 1987. GaAs Devices and Circuits. New York: Plenum Press.

Shur, M., B. Gelment, C. Saavedra-Munoz, and G. Kelner. 1993. Potential of wide bandgap devices for high-temperature applications. Pp. 465-470 in Proceedings of the 5th International Conference on Silicon-Carbide and Related Materials, Washington, D.C., November 1-3. Philadelphia: Institute of Physics Publishing.

Sze, M. 1969. Physics of Semiconductor Devices. New York: John Wiley & Sons.

Trew, R.J. 1994. Personal communication to W.J. Choyke.

Trew, R.J., J. Yan, and P.M. Mock. 1991. The potential of diamond and SiC electronic devices for microwave and millimeter-wave power applications. Proceedings of the IEEE 79(5):598-620.

Vandelin, G.D. 1982. Small-signal and nonlinear applications for GaAs FETs. Pp. 407 in GaAs FET Principles and Technology. J.V. DiLorenzo, ed. Dedham, Massachusetts: Artech House, Inc.

Wemple, S.H., and H. Huang. 1982. Thermal design of power GaAs FETs. Pp. 313 in GaAs FET Principles and Technology. J.V. DiLorenzo, ed. Dedham, Massachusetts: Artech House, Inc.

Appendix D: Biographical Sketches of Committee Members

WOLFGANG J. CHOYKE has been a professor in the Department of Physics at the University of Pittsburgh since 1988. He spent the previous 36 years at the Westinghouse Research Laboratories. He received a B.Sc. and a Ph.D. from Ohio State University.

MICHAEL G. ADLERSTEIN is a principal scientist at the Raytheon Research Division. He received a B.S. and an M.S. from the Polytechnic Institute of Brooklyn and a Ph.D. in applied physics from Harvard University.

JEROME J. CUOMO has been a professor in the Department of Materials Science and Engineering at North Carolina State University since 1993. He was previously senior manager at IBM's Materials Laboratory Center for Science and Services. He received a B.S. from Manhattan College, an M.S. in physical chemistry from St. Johns University, and a Ph.D. in physics from Odense University in Denmark. He is a member of the National Academy of Engineering.

ARTHUR G. FOYT, JR., is manager of Electronics Research at the United Technologies Research Center. He received a B.S., an M.S., and a Sc.D. in electrical engineering from the Massachusetts Institute of Technology.

EVELYN L. HU is professor and chair of the Deparment of Electrical and Computer Engineering at the University of California, Santa Barbara. She received a B.A. from Barnard College and an M.A. and a Ph.D. in physics from Columbia University.

LIONEL C. KIMERLING has been a professor in the Department of Materials Science and Engineering at the Massachusetts Institute of Technology since 1991. He spent the previous 10 years as head of the Materials Physics Research Department at Bell Laboratories. He received an S.B. and a Ph.D. from the Massachusetts Institute of Technology. He is also a member of the National Materials Advisory Board.

MARK R. PINTO is department head of ULSI at AT&T Bell Laboratories. He received a B.S. in electrical engineering and computer science from Rensselaer Polytechnic Institute and an M.S. and a Ph.D. in electrical engineering from Stanford University.

MICHAEL A. TAMOR is staff scientist in the physics department and group leader of the Diamond Film Project at Ford Motor Company. He received a B.S. from the University of California, Los Angeles, and an M.S. and a Ph.D. in physics from the University of Illinois, Urbana.

IWONA TURLIK is vice-president and director of Motorola's Corporate Manufacturing Research Center. She received an M.S. in electrical engineering and a Ph.D. in technical sciences from the Technical University of Wroclaw, Poland.